New York State Coach, Empire Edition, Mathematics, Grade 8

Triumph Learning®

New York State Coach, Empire Edition, Mathematics, Grade 8
316NY
ISBN-10: 1-60824-077-0
ISBN-13: 978-1-60824-077-7

Contributing Writer: John Nelson
Cover Image: The Empire State Building. © Jupiterimages/Comstock

Triumph Learning® 136 Madison Avenue, 7th Floor, New York, NY 10016

Printed in the United States of America.

10 9 8 7 6 5 4 3 2

Table of Contents

* Grade 7 May–June Indicators ** Grade 8 May–June Indicators

* Grade 7 May–June Indicators ** Grade 8 May–June Indicators

Letter to the Student

Dear Student,

Welcome to *Coach*! This book provides instruction and practice that will help you master all the important skills you need to know, and gives you practice answering the kinds of questions you will see on your state's test.

The *Coach* book is organized into chapters and lessons, and includes two Practice Tests. Before you begin the first chapter, your teacher may want you to take Practice Test 1, which will help you identify skill areas that need improvement. Once you and your teacher have identified those skills, you can select the corresponding lessons and start with those. Or, you can begin with the first chapter of the book and work through to the end.

Each of the lessons has three parts. The first part walks you through the skill so you know just what it is and what it means. The second part gives you a model, or example, with hints to help your thinking about the skill. And the third part of the lesson gives you practice with the skill to see how well you understand it.

After you have finished all the lessons in the book, you can take Practice Test 2 to see how much you have improved. And even if you did well on Practice Test 1, you'll probably do better on Practice Test 2 because practice makes perfect!

We wish you lots of success this year, and hope the *Coach* will be a part of it!

Test-Taking Checklist

Here are some tips to keep in mind when taking a test. Take a deep breath. You'll be fine!

- ✓ Follow the directions! Remember, you won't get points if you don't do what the directions say!
- ✓ If you're having trouble understanding a question, try to reword it. How else can the question be asked?
- ✓ On questions you're not sure about, eliminate all answers that you are positive are incorrect. Then choose the answer that seems right.
- ✓ Really stumped? Skip the question and come back to it later.
- ✓ Be extra aware of words that are **bolded**, *italicized*, or <u>underlined</u>. They are usually important.
- ✓ Graphs and charts contain important information. Illustrations often provide clues.
- ✓ If you're allowed, use scrap paper. Take notes or make sketches to help you answer questions.
- ✓ Read all the answer choices before picking the best answer. Sometimes more than one answer may be true. Your job is to choose the best answer.
- ✓ Make sure you've marked your answer correctly. Double-check your answer sheet every ten questions to make sure you're on the right number.
- ✓ If you finish early, read over your answers to check for mistakes. But don't get too caught up in changing your answers—your initial answer is usually correct.
- ✓ Spend a reasonable amount of time on each question. Don't rush through, but make sure to keep up your pace, too. You don't want to run out of time.

Good Luck!

New York State Math Indicators Correlation Chart

*Grade 7 May–June Indicators ** Grade 8 May–June Indicators

Indicator	New York State Grade 8 Math Indicators	*Coach* Lesson(s)
	STRAND 1: NUMBER SENSE AND OPERATIONS	
Operations: Students will understand meanings of operations and procedures, and how they relate to one another.		
8.N.1	Develop and apply the laws of exponents for multiplication and division	1, 10
8.N.2	Evaluate expressions with integral exponents	1
8.N.3	Read, write, and identify percents less than 1% and greater than 100%	2
8.N.4	Apply percents to: tax, percent increase/decrease, simple interest, sale price, commission, interest rates, gratuities	2, 3
Estimation: Students will compute accurately and make reasonable estimates.		
8.N.5	Estimate a percent of quantity, given an application	4
8.N.6	Justify the reasonableness of answers using estimation	4
	STRAND 2: ALGEBRA	
Variables and Expressions: Students will represent and analyze algebraically a wide variety of problem solving situations.		
8.A.1	Translate verbal sentences into algebraic inequalities	6
8.A.2	Write verbal expressions that match given mathematical expressions	5
8.A.3	Describe a situation involving relationships that matches a given graph	8
8.A.4	Create a graph given a description or an expression for a situation involving a linear or nonlinear relationship	8
8.A.5	Use physical models to perform operations with polynomials	9, 10
Variables and Expressions: Students will perform algebraic procedures accurately.		
8.A.6	Multiply and divide monomials	10
8.A.7	Add and subtract polynomials (integer coefficients)	9
8.A.8	Multiply a binomial by a monomial or a binomial (integer coefficients)	11
8.A.9	Divide a polynomial by a monomial (integer coefficients) *Note: The degree of the denominator is less than or equal to the degree of the numerator for all variables.*	10
8.A.10	Factor algebraic expressions using the GCF	12
8.A.11	Factor a trinomial in the form $ax^2 + bx + c$; $a = 1$ and c having no more than three sets of factors	13
Equations and Inequalities		
8.A.12	Apply algebra to determine the measure of angles formed by or contained in parallel lines cut by a transversal and by intersecting lines	18, 19
8.A.13	Solve multi-step inequalities and graph the solution set on a number line	7
8.A.14	Solve linear inequalities by combining like terms, using the distributive property, or moving variables to one side of the inequality (include multiplication or division of inequalities by a negative number)	7

*Grade 7 May–June Indicators ** Grade 8 May–June Indicators

Indicator	New York State Grade 8 Math Indicators	*Coach* Lesson(s)
Patterns, Relations, and Functions: Students will recognize, use, and represent algebraically patterns, relations, and functions.		
*7.A.9	Build a pattern to develop a rule for determining the sum of the interior angles of polygons	15
*7.A.10	Write an equation to represent a function from a table of values	15
8.A.15	Understand that numerical information can be represented in multiple ways: arithmetically, algebraically, and graphically	14
8.A.16	Find a set of ordered pairs to satisfy a given linear numerical pattern (expressed algebraically); then plot the ordered pairs and draw the line	14
**8.A.17	Define and use correct terminology when referring to function (domain and range)	17
**8.A.18	Determine if a relation is a function	17
**8.A.19	Interpret multiple representations using equation, table of values, and graph	16
STRAND 3: GEOMETRY		
Constructions: Students will use visualization and spatial reasoning to analyze characteristics and properties of geometric shapes.		
**8.G.0	Construct the following, using a straight edge and compass: Segment congruent to a segment Angle congruent to an angle Perpendicular bisector Angle bisector	22
Geometric Relationships: Students will identify and justify geometric relationships, formally and informally.		
8.G.1	Identify pairs of vertical angles as congruent	18
8.G.2	Identify pairs of supplementary and complementary angles	18
8.G.3	Calculate the missing angle in a supplementary or complementary pair	18
8.G.4	Determine angle pair relationships when given two parallel lines cut by a transversal	19
8.G.5	Calculate the missing angle measurements when given two parallel lines cut by a transversal	19
8.G.6	Calculate the missing angle measurements when given two intersecting lines and an angle	18
Transformational Geometry: Students will apply transformations and symmetry to analyze problem solving situations.		
8.G.7	Describe and identify transformations in the plane, using proper function notation (rotations, reflections, translations, and dilations)	20, 21
8.G.8	Draw the image of a figure under rotations of 90 and 180 degrees	20
8.G.9	Draw the image of a figure under a reflection over a given line	20
8.G.10	Draw the image of a figure under a translation	20
8.G.11	Draw the image of a figure under a dilation	21
8.G.12	Identify the properties preserved and not preserved under a reflection, rotation, translation, and dilation	20, 21

*Grade 7 May–June Indicators ** Grade 8 May–June Indicators

Indicator	New York State Grade 8 Math Indicators	*Coach* Lesson(s)
Coordinate Geometry: Students will apply coordinate geometry to analyze problem solving situations.		
8.G.13	Determine the slope of a line from a graph and explain the meaning of slope as a constant rate of change	23
8.G.14	Determine the y-intercept of a line from a graph and be able to explain the y-intercept	24
8.G.15	Graph a line using a table of values	24
8.G.16	Determine the equation of a line given the slope and the y-intercept	25
8.G.17	Graph a line from an equation in slope-intercept form ($y = mx + b$)	25
8.G.18	Solve systems of equations graphically (only linear, integral solutions, $y = mx + b$ format, no vertical/horizontal lines)	26
8.G.19	Graph the solution set of an inequality on a number line	7
8.G.20	Distinguish between linear and nonlinear equations $ax^2 + bx + c; a = 1$ (only graphically)	27
8.G.21	Recognize the characteristics of quadratics in tables, graphs, equations, and situations	28
STRAND 4: MEASUREMENT		
Units of Measurement: Students will determine what can be measured and how, using appropriate methods and formulas.		
*7.M.1	Calculate distance using a map scale	33
*7.M.5	Calculate unit price using proportions	32
*7.M.6	Compare unit prices	32
*7.M.7	Convert money between different currencies with the use of an exchange rate table and a calculator	31
8.M.1	Solve equations/proportions to convert to equivalent measurements within metric and customary measurement systems *Note: Also allow Fahrenheit to Celsius and vice versa.*	29, 30

STRAND

Number Sense and Operations

Lesson

1 Evaluate Expressions with Exponents

8.N.1, 8.N.2

Getting the Idea

Exponents are shortcuts for recording repeated multiplication. A number written in **exponential form** contains a **base** and an exponent. The exponent tells how many times the base is used as a factor.

Numbers written in exponential form can also be written in **standard form**, which is a way of writing a number using only digits.

$7^2 = 49$ 7 is the base.

2 is the exponent.

49 is the number in standard form.

7^2 is a way to write 49 in exponential form. 7^2 can also be called a power of 7, and is read as "7 to the 2nd power."

EXAMPLE 1

What is 4^3 written in standard form?

STRATEGY **Use the base as a factor the number of times indicated by the exponent.**

4 is the base.

3 is the exponent.

Use 4 as a factor 3 times.

$4 \cdot 4 \cdot 4 = 64$

Remember that $4 \cdot 4 \cdot 4$ is the same as $4 \times 4 \times 4$.

SOLUTION **4^3, written in standard form, is 64.**

A power with a negative exponent is equivalent to the reciprocal of the power: $7^{-2} = \frac{1}{7^2}$. If an expression or number in a numerator has a negative exponent, move the expression or number to the denominator and change the sign of the exponent. If an expression or number in a denominator has a negative exponent, move the expression or number to the numerator and change the sign of the exponent. Look at the examples below.

$$\frac{5^{-3}}{2} = \frac{2}{5^3} \qquad \frac{1}{6^{-2}} = \frac{6^2}{1} = 6^2$$

EXAMPLE 2

What is 8^{-3} written in standard form?

STRATEGY **Move the expression to the denominator of a fraction and change the sign of the exponent. Then write the expression in standard form.**

STEP 1 Move 8^{-3} to the denominator of a fraction and change the sign of the exponent.

$$8^{-3} = \frac{8^{-3}}{1} = \frac{1}{8^3}$$

STEP 2 Write the expression in standard form.

$$\frac{1}{8^3} = \frac{1}{8 \times 8 \times 8} = \frac{1}{512}$$

SOLUTION $\mathbf{8^{-3} = \frac{1}{512}}$

EXAMPLE 3

What is $\frac{1}{2^{-4}}$ written in standard form?

STRATEGY **Move the expression to the numerator of a fraction and change the sign of the exponent. Then write the expression in standard form.**

STEP 1 Move 2^{-4} to the numerator of a fraction and change the sign of the exponent.

$$\frac{1}{2^{-4}} = \frac{2^4}{1} = 2^4$$

STEP 2 Write the expression in standard form.

$$2^4 = 2 \cdot 2 \cdot 2 \cdot 2 = 16$$

SOLUTION $\mathbf{\frac{1}{2^{-4}} = 16}$

Sometimes, you may need to compute with exponents. The exponent rules in the table below can help you.

Exponent Rule	Examples
To multiply two powers with the same base, add the exponents.	$a^s \times a^t = a^{s+t}$ $8^2 \times 8^7 = 8^{2+7} = 8^9$
To divide two powers with the same base, subtract the exponents.	$a^s \div a^t = a^{s-t}$ $3^{10} \div 3^2 = 3^{10-2} = 3^8$
To raise a power to a power, multiply the exponents.	$(a^s)^t = a^{s \times t}$ $(6^4)^5 = 6^{4 \times 5} = 6^{20}$

EXAMPLE 4

What is $3^2 \cdot 3^4$? Write the product in standard form.

STRATEGY **Use the appropriate exponent rule and then evaluate.**

STEP 1 Identify the operation.

The operation is multiplication and the exponents have the same base, so add the exponents.

$3^2 \cdot 3^4 = 3^{2+4} = 3^6$

STEP 2 Evaluate the expression to write it in standard form.

$3^6 = 3 \cdot 3 \cdot 3 \cdot 3 \cdot 3 \cdot 3 = 729$

SOLUTION $\mathbf{3^2 \cdot 3^4 = 3^6 = 729}$

Note: You could have solved the problem in Example 4 by evaluating each factor and then multiplying.

$3^2 = 9$ and $3^4 = 81$, so $3^2 \cdot 3^4 = 9 \cdot 81 = 729$.

EXAMPLE 5

What is $5^5 \div 5^3$? Write the quotient in standard form.

STRATEGY **Use the appropriate exponent rule and then evaluate.**

STEP 1 Identify the operation.

The operation is division and the exponents have the same base, so subtract the exponents.

$5^5 \div 5^3 = 5^{5-3} = 5^2$

STEP 2 Evaluate the expression and write it in standard form.

$5^2 = 5 \times 5 = 25$

SOLUTION $\mathbf{5^5 \div 5^3 = 5^2 = 25}$

EXAMPLE 6

What is $7^4 \times 7^{-2}$ written in standard form?

STRATEGY **Use the appropriate exponent rule and then evaluate.**

STEP 1 Identify the operation.

The operation is multiplication and the exponents have the same base, so add the exponents.

$7^4 \times 7^{-2} = 7^{4+(-2)} = 7^2$

STEP 2 Evaluate the expression and write it in standard form.

$7^2 = 7 \times 7 = 49$

SOLUTION $\mathbf{7^4 \times 7^{-2} = 7^2 = 49}$

COACHED EXAMPLE

Write the product $(3^2)^3 \times 3^3$ in exponential form and in standard form.

THINKING IT THROUGH

Do both numbers have the same base? _____________

Use exponent rules to write the number in exponential form.

To raise a power to a power, I should _____________ the exponents.

$(3^2)^3 =$ ___________________

To multiply two numbers with the same base, I should _____________ the exponents.

_____ $\times 3^3 =$ ___________________

In exponential form, the exponent of the number is ________. That means that the base, __________, is used as a factor __________ times.

Multiply to find the standard form of the number. ____

In exponential form, the product is ____________.

In standard form, the product is _________________.

Lesson Practice

Choose the correct answer.

1. What is 5^4 in standard form?

A. 20

B. 625

C. 1,025

D. 3,125

2. Which shows 9^{-3} in standard form?

A. 729

B. 27

C. $\frac{1}{27}$

D. $\frac{1}{729}$

3. Which shows $6^3 \times 6^4$ in exponential form?

A. 6^7

B. 6^{12}

C. 36^7

D. 36^{12}

4. What is $4^4 \cdot 4^2$ in standard form?

A. 64

B. 128

C. 4,096

D. 65,536

5. What is $6^{-1} \div 6^4$ in exponential form?

A. 6^{-5}

B. 6^{-3}

C. 6^1

D. 6^3

6. Which shows $4^6 \div 4^5$ in standard form?

A. 0

B. 1

C. 4

D. 16

7. What is $(11^6)^2$ in exponential form?

A. 22^6

B. 11^{12}

C. 11^8

D. 11^4

8. What is $(5^{-2})^3$ in standard form?

A. $\frac{1}{15{,}625}$

B. $\frac{1}{3{,}125}$

C. $\frac{1}{5}$

D. 5

9. Simplify: $4 \times 2^6 \div 2^3$

Answer ______________

10. Simplify: $(2^4)^3 \times 2^{-5}$

Answer ____________

Lesson

2 Percents

8.N.3, 8.N.4

Getting the Idea

A **percent** (%) is a ratio that compares a number to 100. You can write the ratio whose numerator is 56 and whose denominator is 100 as 56% and read it as "fifty-six percent."

Because a percent is a ratio, you can write a percent as a fraction.

$56\% = \frac{56}{100} = \frac{14}{25}$

You can also express a percent as a decimal by dividing the percent by 100 and dropping the percent symbol.

$56\% = 56 \div 100 = 0.56$

A shortcut for expressing a percent as a decimal is to move the decimal point two places to the left and drop the percent symbol. Remember that a percent with an integer value can be renamed to have a decimal point: 56% = 56.0%. Moving the decimal point two places to the left and dropping the percent symbol results in 0.560, or 0.56.

Use the same methods for writing a percent as a fraction and as a decimal when the percent is less than 1%.

$\frac{1}{2}\% = \frac{\frac{1}{2}}{100} = \frac{1}{2} \times \frac{1}{100} = \frac{1}{200}$

$\frac{1}{2}\% = 0.5\% = 0.005$

Use the same methods for writing a percent as a fraction and as a decimal when the percent is greater than 100%.

$160\% = \frac{160}{100} = \frac{16}{10} = \frac{8}{5}$, or $1\frac{3}{5}$

$160\% = 160.0\% = 1.600$, or 1.6

To express a decimal as a percent, multiply the decimal by 100 and write the percent symbol next to the result. A shortcut is to move the decimal point two places to the right and write the percent symbol next to the result.

$1.32 = 1.32 \times 100 = 132\%$

To express a fraction as a percent, first write the fraction as a decimal by dividing the numerator by the denominator. Then express the decimal as a percent.

$\frac{7}{40} = 7 \div 40 = 0.175$

$0.175 = 17.5\%$

You can identify percents from models. The shaded portion of the 10-by-10 grid below represents 37%.

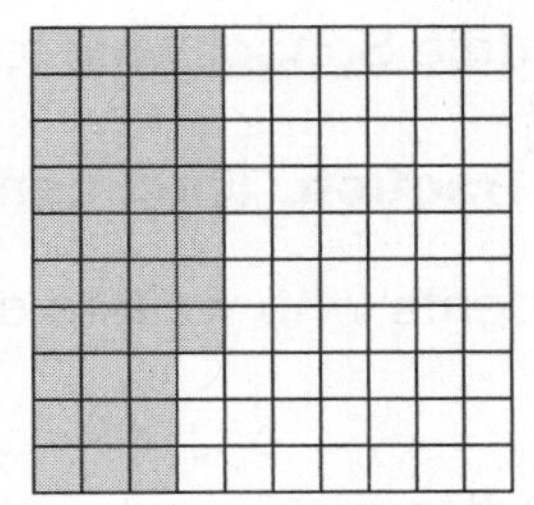

EXAMPLE 1

Each circle represents 100%. What percent do the shaded portions represent?

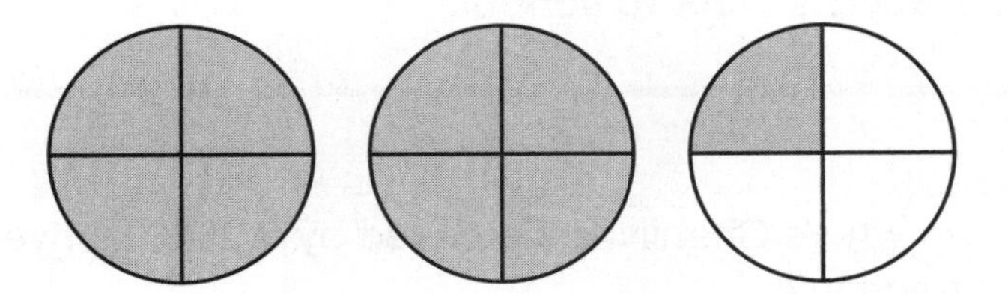

STRATEGY **Determine the percent represented by the first two circles. Then find the percent of the third circle that is shaded.**

STEP 1 Determine the percent represented by the first two circles.

The first two circles are completely shaded, so they represent $100\% + 100\%$, or 200%.

STEP 2 Find the percent of the third circle that is shaded.

One fourth of the circle is shaded. Write this as a percent.

$\frac{1}{4} = 1 \div 4 = 0.25 = 25\%$

STEP 3 Combine the results of Steps 1 and 2.

$200\% + 25\% = 225\%$

SOLUTION **The shaded portions represent 225%.**

EXAMPLE 2

There are 250 students at Williams Middle School. Of those students, 84 walk to school. What percent of the students at Williams Middle School walk to school?

STRATEGY **Write the number as a fraction. Then convert to a percent.**

STEP 1 Write the fraction of students who walk to school.

$\frac{84}{250}$

STEP 2 Convert the fraction to a decimal.

$\frac{84}{250} = 84 \div 250 = 0.336$

STEP 3 Convert the decimal to a percent.

$0.336 = 33.6\%$

SOLUTION **33.6% of students walk to school.**

EXAMPLE 3

The population of the town where Chet lives increased by 0.25% last year. What is this increase expressed as a fraction?

STRATEGY **Write the percent as a decimal. Then convert to a fraction.**

STEP 1 Write the percent as a decimal.

Divide the percent by 100.

$0.25\% = 0.25 \div 100 = 0.0025$

STEP 2 Convert the decimal to a fraction.

$0.0025 = \frac{25}{10,000} = \frac{1}{400}$

SOLUTION **The increase of 0.25%, expressed as a fraction, is $\frac{1}{400}$.**

EXAMPLE 4

A company reported that its 2008 profit was $3\frac{4}{5}$ times its 2007 profit. What is $3\frac{4}{5}$ expressed as a percent?

STRATEGY **Write the mixed number as a decimal. Then convert the decimal to a percent.**

STEP 1 Convert the fraction in the mixed number to a decimal.

$$3\frac{4}{5} = 3 + \frac{4}{5} = 3 + (4 \div 5) = 3 + 0.8 = 3.8$$

STEP 2 Convert the decimal to a percent.

$$3.8 = 380\%$$

SOLUTION **Expressed as a percent, $3\frac{4}{5}$ is 380%.**

COACHED EXAMPLE

What is $\frac{3}{500}$ expressed as a percent? What is 464% expressed as a mixed number?

THINKING IT THROUGH

Express $\frac{3}{500}$ as a percent.

First, convert $\frac{3}{500}$ to a decimal by dividing ____________ by ____________.

Expressed as a decimal, $\frac{3}{500}$ is ____________.

Then rename the decimal form of $\frac{3}{500}$ as a percent by moving the decimal point ____________ places to the ____________. Add the percent sign.

The fraction $\frac{3}{500}$, expressed as a percent, is ____________.

Express 464% as a mixed number.

First, convert 464% to a decimal by moving the decimal point ____________ places to the ____________.

Expressed as a decimal, 464% is ____________.

Then convert the decimal form of 464% to a mixed number.

The whole number part of the mixed number is ____________.

The fraction part of the mixed number is ____________.

In simplest form, the fraction part is ____________.

The percent 464%, expressed as a mixed number, is ____________.

Lesson Practice

Choose the correct answer.

1. A percent is modeled below.

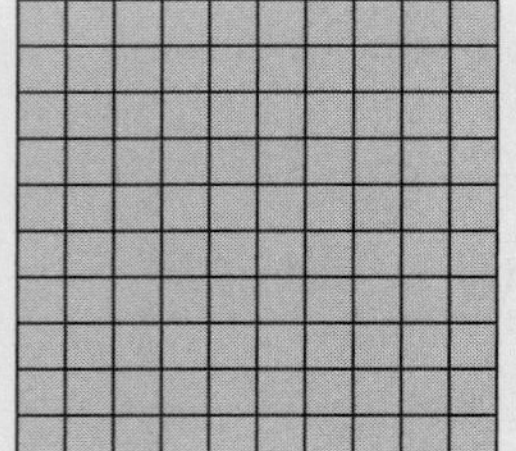
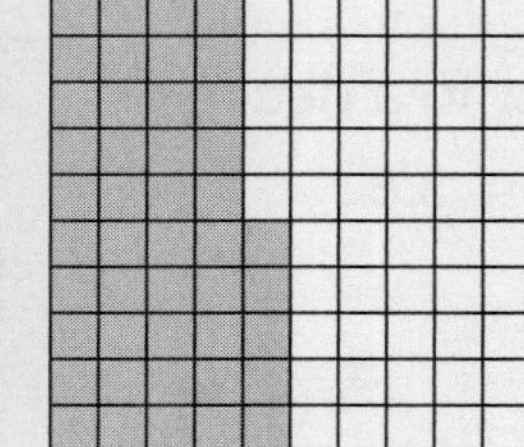

What percent does the shaded portion of the model represent?

A. 45%
B. 55%
C. 145%
D. 155%

2. What is 3 out of 10 expressed as a percent?

A. 0.3%
B. 3%
C. 30%
D. 300%

3. What is 3.5% expressed as a decimal?

A. 0.0035
B. 0.035
C. 0.35
D. 3.5

4. What percent of 250 is 76?

A. 30.4%
B. 34.6%
C. 36.6%
D. 40.4%

5. A novel has 640 pages. Jorge read 4 pages of the novel during his lunch period. What percent of the novel did Jorge read during his lunch period?

A. 0.00625%
B. 0.0625%
C. 0.625%
D. 6.25%

6. Gladys figures that on average she spends 28% of each day sleeping. What fraction of a day does she sleep?

A. $\frac{1}{28}$
B. $\frac{7}{25}$
C. $\frac{7}{20}$
D. $\frac{6}{7}$

7. Kendra increased the number of stamps in her collection by 175%. What is 175% expressed as a mixed number?

Answer ________________

8. Mitch estimates that his chance of winning a raffle prize is 1 in 800. What is his chance of winning, expressed as a percent?

Answer ________________

EXTENDED-RESPONSE QUESTION

9. Ruiz is reading a book for his social studies class. So far, he has read 7.5% of the book.

Part A What is this percent expressed as a decimal? What is this percent expressed as a fraction?

Part B If the book is 480 pages long, how many pages has Ruiz read so far?

Lesson

3 Applications of Percents

8.N.4

Getting the Idea

Percents have a wide variety of applications. The media (radio, television, newspapers, the Internet, and so on) are filled with information expressed in terms of percents, so it is important to understand how percents are applied. Percents can express or determine an amount of increase or decrease. In situations involving money, applications of percents include finding amounts of interest and interest rates, taxes and tax rates, and commissions.

Sales tax is the amount of money the government charges on the sale of an item or service. That amount is paid in addition to the cost of the item or service.

EXAMPLE 1

Jordan bought a computer with a price of $775.50. The sales tax rate on the computer was 8%. How much did Jordan pay for the computer, including sales tax?

STRATEGY **Write the percent as a decimal and multiply to find the amount of tax. Then add the tax to the price of the item.**

STEP 1 Write 8% as a decimal.

$8\% = 8 \div 100 = 0.08$

STEP 2 Multiply the price of the computer by the decimal.

$775.50 \times 0.08 = 62.04$

STEP 3 Add the price of the computer and the sales tax.

$775.50 + 62.04 = 837.54$

SOLUTION **Jordan paid $837.54, including sales tax, for the computer.**

A **discount** is an amount of money that is taken off the original price of an item or service. The **sale price** is the cost of the item after the discount has been applied.

EXAMPLE 2

Haley bought a CD at 20% off. The CD originally cost \$17.75. What is the sale price of the CD?

STRATEGY **Write the percent as a decimal and multiply to find the amount of discount. Then subtract the discount from the original cost.**

STEP 1 Write the percent as a decimal.

$20\% = 0.2$

STEP 2 Multiply to find the amount of discount.

$17.75 \times 0.2 = 3.55$

The discount is \$3.55.

STEP 3 Subtract.

$17.75 - 3.55 = 14.20$

SOLUTION **The sale price of the CD is \$14.20.**

Here is another way to solve Example 2. The discount is 20%, so the sale price will be 100% (the original cost) minus 20%, or 80% of the original cost. Multiply the original cost by 80% in decimal form: $17.75 \times 0.8 = 14.20$

Simple interest is money paid or earned based on the **principal** (the amount of money borrowed or saved). To calculate simple interest, use the formula $I = prt$, where I is the interest amount, p is the principal, r is the **interest rate**, and t is the time (in years).

EXAMPLE 3

Daniel has \$1,200 in his savings account. He receives 0.75% annual simple interest. If he makes no further deposits or withdrawals, how much money will he earn in simple interest in 2 years?

STRATEGY **Use the formula $I = prt$.**

STEP 1 Convert the percent to a decimal.

$0.75\% = 0.0075$

STEP 2 Multiply.

$I = p \times r \times t$

$I = 1{,}200 \times 0.0075 \times 2 = 18$

SOLUTION **Daniel will earn \$18 in simple interest in 2 years.**

A **commission** is an amount of money earned by a salesperson, based on a percent of total sales.

EXAMPLE 4

Ms. Harris sold a house for $200,000. She received $5,500 in commission for the sale of the house. What is the rate of commission that Ms. Harris received? Write the answer as a percent.

STRATEGY **Divide the commission by the amount of the sale.**

STEP 1 Divide the commission by the amount of the sale.

$$\frac{5{,}500}{200{,}000} = 5{,}500 \div 200{,}000 = 0.0275$$

STEP 2 Convert the decimal to a percent.

$$0.0275 = 2.75\%$$

SOLUTION **Ms. Harris received a 2.75% commission on the sale of the house.**

To find a **percent of increase** or a **percent of decrease**, first write the difference of the original amount and the new amount as the numerator of a fraction. Then write the original amount as the denominator. Convert the fraction to a percent.

EXAMPLE 5

A basketball team won 15 games during the 2007 season. In the 2008 season, the team won 21 games. What was the percent of increase in the number of games the team won?

STRATEGY **Write a fraction with the difference of the original amount and the new amount as the numerator over the original amount as the denominator. Then convert the fraction to a percent.**

STEP 1 Find the difference of the number of games won in 2007 and in 2008.

$$21 - 15 = 6$$

STEP 2 Write a fraction with the difference as the numerator and the number of games won in the 2007 season as the denominator.

$$\frac{6}{15}$$

STEP 3 Convert the fraction to a percent.

You can simplify the fraction first.

$$\frac{6}{15} = \frac{6 \div 3}{15 \div 3} = \frac{2}{5} = 40\%$$

SOLUTION **The percent of increase in the number of games won was 40%.**

COACHED EXAMPLE

The Hayes family went to dinner at a restaurant. The check came to $68.24 plus tax. The tax rate was 8.5%. They decided to leave a 15% tip based on the total cost of dinner, including tax. How much did they pay in all?

THINKING IT THROUGH

Multiply the check amount by the tax rate to find the total cost of dinner.

Convert 8.5% to a decimal.

8.5% = ____________

Multiply to find the amount of tax.

____________ × 68.24 = ____________

Rounded to the nearest cent, the tax is ____________.

The total cost of dinner before tip is ____________ + ____________ = ____________

To find the tip amount, multiply the total cost of dinner by 15%.

Convert 15% to a decimal.

15% = ____________

Multiply the total cost by that decimal.

____________ × ____________ = ____________

Rounded to the nearest cent, the tip is ____________.

Add the tip to the total cost of the dinner.

____________ + ____________ = ____________

The Hayes family paid ____________ in all for dinner.

Lesson Practice

Choose the correct answer.

1. Chase treated his friends to lunch, which cost $33.35 before the tip. He left a 15% tip. How much money did he leave as a tip?

A. $0.50

B. $1.67

C. $5.00

D. $6.67

2. The 30%-off sale at Just Books is this week. Vidisha wants to buy a book that regularly costs $16.40. How much will the book cost on sale?

A. $4.92

B. $11.48

C. $16.10

D. $21.32

3. Greta deposited $800 in a savings account that pays 0.95% annual simple interest. If she leaves the money in this account for 4 years, how much interest will Greta earn?

A. $3.40

B. $7.60

C. $30.40

D. $304.00

4. Juliet bought a pair of pants that normally cost $40 for a sale price of $24. What was the percent discount on the pants?

A. 16%

B. 40%

C. 60%

D. $66\frac{2}{3}$%

5. Ms. Sinco earns an annual salary of $60,000 plus a commission of 6% on all sales that she makes. If she made $750,000 in sales last year, how much money did she earn in all?

A. $45,000

B. $63,600

C. $64,500

D. $105,000

6. Carlos bought a flat-screen television for $1,100. Six months later, he bought the same model of television for his mother and paid $800. What was the percent of decrease in the price of the television, rounded to the nearest tenth of a percent?

A. 27.3%

B. 30.0%

C. 37.5%

D. 73.7%

7. The value of Pam's house has increased by 38% since she bought it. If Pam paid $175,000 for her house, what is the current value of Pam's house?

Answer ______________

8. James wants a DVD player that is advertised at $124.99. How much would he save if he buys the DVD player in a county with a sales tax rate of 7.25% compared to a county with a sales tax rate of 8.25%?

Answer ______________

EXTENDED-RESPONSE QUESTION

9. Colleen works as a real estate agent. She makes 4.25% in commission for every house she sells.

Part A She made $9,350 in commission for the last house she sold. How much did that house cost?

Part B Explain how you know your answer is correct.

____

____

____

Lesson

4 Estimation with Percents

Getting the Idea

An **estimate** is a rough calculation or an approximation, not an exact answer. When you work with percents, knowing how to estimate can help you either to find a good approximation of an answer or to check the reasonableness of an answer.

One way to estimate is to use rounding. In general, to round a number, look at the digit directly to the right of the place to which you are rounding.

- If that digit is less than or equal to 4, round down.
- If that digit is greater than or equal to 5, round up.

EXAMPLE 1

André's bill at a restaurant is $7.73. He wants to leave a 15% gratuity. (A gratuity is the same as a tip.) How much should André leave for the gratuity if he rounds the bill to the nearest dollar?

STRATEGY **Use rounding.**

STEP 1 Round the amount of the bill to the nearest dollar.

$7.73 rounded to the nearest dollar is $8.

STEP 2 Multiply to find 15% of $8.

Convert 15% to a decimal.

15% of $8 = 0.15 \times 8 = 1.20$

SOLUTION **André should leave $1.20 for the gratuity.**

EXAMPLE 2

Ms. Davis earns an annual salary of $63,825. The tax rate on her income is 28%. About how much money does Ms. Davis pay in taxes each year?

STRATEGY **Round each number. Then compute.**

STEP 1 Round Ms. Davis's salary to the nearest $10,000.

In $63,825, the digit to the right of the ten thousands place is 3, so round down to $60,000.

STEP 2 Round the percent to the nearest ten percent.

Since $8 \geq 5$, round 28% up to 30%.

STEP 3 Multiply the rounded numbers.

$30\% \text{ of } 60{,}000 = 0.3 \times 60{,}000 = 18{,}000$

SOLUTION **Ms. Davis pays about $18,000 in taxes each year.**

EXAMPLE 3

In 1990 the population of Dobbs Ferry, New York, was 9,940. By 2004 the population had grown to 16,003. Victor claims that Dobbs Ferry's population increased by 38% from 1990 to 2004. Is Victor's claim reasonable?

STRATEGY **Round the numbers and estimate the percent of increase.**

STEP 1 Round the numbers to the nearest thousand.

9,940 rounds to 10,000.

16,003 rounds to 16,000.

STEP 2 Find the percent of increase.

Find the difference of the populations.

$16{,}000 - 10{,}000 = 6{,}000$

Write a fraction with the population difference in the numerator and the original population in the denominator. Then rename the fraction as a percent.

$\frac{6{,}000}{10{,}000} = \frac{6}{10} = 0.6 = 60\%$

SOLUTION **The estimated population increase of 60% is not close to Victor's claim of 38%. Victor's claim is not reasonable.**

COACHED EXAMPLE

Clarence's salary will increase by 3.24% next year. His current salary is $58,775. What is a good estimate of his salary for next year?

THINKING IT THROUGH

Round the percent of increase to the nearest whole percent.

To the nearest whole percent, 3.24% rounds to ____________.

Round Clarence's current salary to the nearest thousand.

To the nearest thousand, $58,775 rounds to ____________.

Multiply the estimated percent and estimated salary.

____

Add the estimated increase to the estimate of his current salary.

____

A good estimate of Clarence's salary for next year is ____________.

Lesson Practice

Choose the correct answer.

1. A skateboard costs $52.89. The sales tax rate is 7%. About how much is the sales tax on the skateboard?

   A. $3.50
   B. $7.00
   C. $10.50
   D. $35.00

2. Leon's restaurant bill came to $28.64, including tax. He wants to leave an 18% tip on the total. About how much should he leave for a tip?

   A. $4.00
   B. $6.00
   C. $8.00
   D. $10.00

3. Lily works in an appliance store. She earns a monthly base salary of $2,000 plus a 5.25% commission on her sales. Last month she sold $4,975 worth of appliances. Which is the best estimate of how much Lily earned last month?

   A. $2,150
   B. $2,250
   C. $2,350
   D. $2,450

4. In the last election for city council, 57.9% of the 2,942 eligible voters cast votes. Which is closest to the number of eligible voters who did **not** vote?

   A. 1,200
   B. 1,400
   C. 1,500
   D. 1,800

5. A tennis racket that regularly sells for $78.25 is on sale for 30% off. Which is the best estimate of the sale price of the racket?

   A. $25
   B. $30
   C. $55
   D. $65

6. Emily borrowed $9,375 to buy a car. The loan carries a simple interest rate of 8.75%. If she makes no payments for the first year, about how much will Emily owe in interest at the end of the year?

   A. $450
   B. $900
   C. $9,450
   D. $9,900

7. There were 14,811 seats sold for a baseball game. The baseball stadium has 44,376 seats. Which is the best estimate of the percent of seats sold?

A. 33%
B. 45%
C. 60%
D. 75%

8. A piano is on sale for $1,695. The sales tax rate is 8.125%. Approximately what is the total cost of the piano?

A. $17
B. $170
C. $1,770
D. $1,870

EXTENDED-RESPONSE QUESTION

9. Last night, the paid attendance at a play was 784 people. The theater manager said that 25% of the audience paid a student rate for their tickets. He determined that 196 people paid the student rate.

Part A Is the theater manager's calculation reasonable?

Part B How did you determine your answer in Part A?

1 Review

1 Which shows $9^{-4} \times 9^{9}$ in exponential form?

A 9^{13}

B 9^{5}

C 9^{-5}

D 9^{-13}

2 Simplify: $5 \times 2^{6} - 6^{3}$

A 42

B 104

C 320

D 999,784

3 What percent does the shaded portion of the model represent?

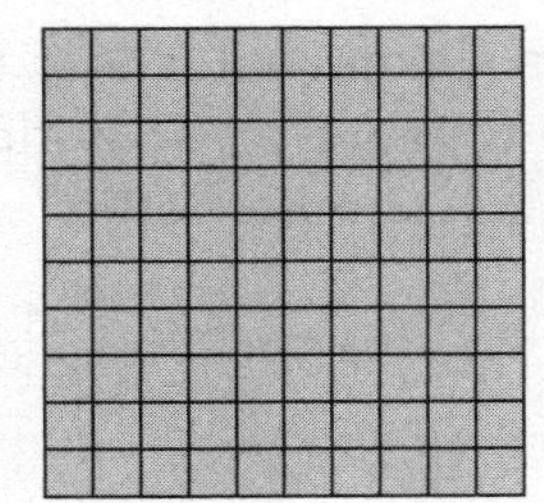

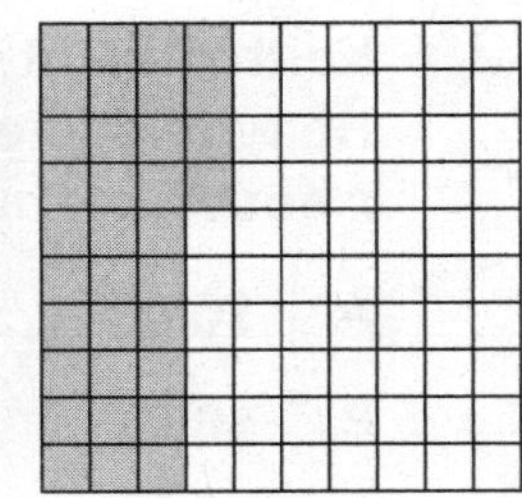

A 34%

B 134%

C 143%

D 166%

4 Fred deposited $1,200 into a savings account that pays 2.5% simple interest annually. If he makes no deposits or withdrawals, how much money will Fred have in his account after 1 year?

A $1,225

B $1,230

C $1,250

D $1,260

5 Emil saw a shirt that he wants to buy on sale for 35% off. If the shirt regularly costs $45, how much will Emil pay for the shirt?

A $29.25

B $30.00

C $31.50

D $44.65

6 A cutting board costs $61.89. The sales tax is 6%. **About** how much is the sales tax on the cutting board?

A $4.00

B $5.00

C $6.00

D $7.00

7 There were 585 students in this year's high school graduating class. The principal reported that 80% of the graduates would be attending college. Nia determined that 468 graduates would attend college. Which statement **best** explains whether Nia's number is reasonable or not, and why?

A No, it is not reasonable because $0.8 \times 500 = 400$.

B No, it is not reasonable because 20% of 600 is 120.

C Yes, it is reasonable because 80% of 730 is 584.

D Yes, it is reasonable because $0.8 \times 600 = 480$.

8 Which shows $3^{-1} \div 3^{4}$ in standard form?

A $\frac{1}{243}$

B $\frac{1}{27}$

C 27

D 243

9 There are 3 green lentils in a bag of brown lentils. If there are 6,000 lentils altogether in the bag, what percent of the lentils are green?

A 0.0005%

B 0.005%

C 0.05%

D 0.5%

10 Austin deposited $550 into a bank account that earns 3% simple interest each year. If he makes no deposits or withdrawals, how much interest will Austin earn in 1 year?

A $1.65

B $16.50

C $165.00

D $551.65

11 Connie is a real estate agent. She earns a base salary of $1,500 per month plus a commission of 2.75% of her sales. Last month her sales totaled $540,000. What were her total earnings last month?

A $13,350

B $14,850

C $14,890

D $16,350

12 Keith's bill at a restaurant is $62.19. He would like to leave an 18% tip. **About** how much should he tip?

A $6.00

B $8.00

C $12.00

D $16.00

13 Last year, 32 eighth-grade students made the honor roll. This year, 36 eighth-grade students made the honor roll. What is the percent of increase in honor roll students from last year to this year?

Show your work.

Answer ______________

14 Simplify the expression below.

$5^3 + 5^4 \div 5^2$

Show your work.

Answer ______________

15 Gary is buying a leather jacket that has a regular price of $250, but is being marked down by 20%.

Part A

What is the sale price of the jacket?

Part B

When Gary brings the jacket to the register, he is given another 15% off the sale price. What will Gary pay for the jacket if there is an 8% sales tax rate? Show your work.

STRAND

2 Algebra

* Grade 7 May—June Indicators ** Grade 8 May—June Indicators

Lesson

5 Write Expressions

Getting the Idea

An **expression** is a mathematical representation that may contain numbers, variables, and operation symbols. Expressions do not include an equality or inequality symbol.

Parts of an expression have special names. Consider this expression: $2x + 3$.

- The variable is x. A **variable** is a symbol used to represent a number.
- The constant is 3. A **constant** is a stand-alone quantity that does not change value.
- The coefficient of x is 2. A **coefficient** is a number that multiplies a variable.

You can write expressions to represent real-world situations. Look for key words that can help you translate the situation into an expression. Here are some examples.

Problem Situation	Key Words	Expression
a charge of $2 plus an additional $0.50 per mile	plus, additional	$2 + 0.5m$
39% of the population	of	$0.39p$
three fewer than twice the number of students	fewer than, twice	$2s - 3$
a number of tiles shared equally by 4 people	shared equally	$n \div 4$

EXAMPLE 1

The money in Leah's bank account is $50 less than half the money in Maggie's account. Let m represent the money in Maggie's account. Write an expression to show the amount of money in Leah's account.

STRATEGY **Use key words to translate the problem situation into an expression.**

STEP 1 What are the key words and what do they represent?

"Less than" indicates subtraction. Note that sometimes these words stand for an inequality symbol ($<$). In this case, though, you are asked to write an expression. "Half" indicates division by 2 $\left(\text{or multiplication by } \frac{1}{2}\right)$.

STEP 2 Translate the words into an expression.

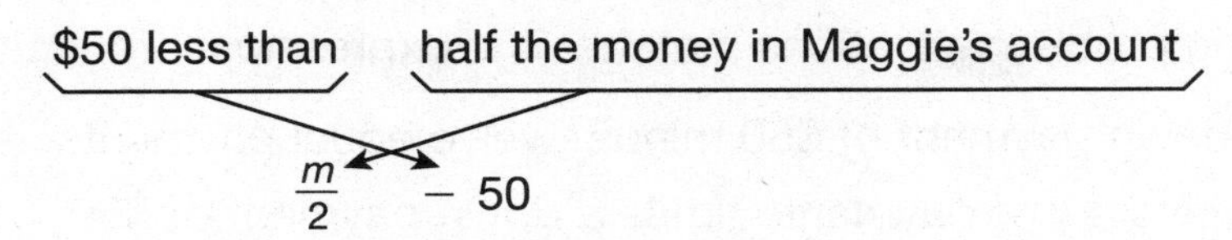

SOLUTION **The amount of money in Leah's account can be expressed as $\frac{m}{2} - 50$.**

EXAMPLE 2

Write a verbal expression that matches the algebraic expression $2n + 100$.

STRATEGY **Use the meaning of each term in the algebraic expression to write a verbal expression.**

STEP 1 What is the meaning of $2n$?

The number 2 is the coefficient of the variable n. This tells you that the unknown quantity n is being multiplied by 2. The term $2n$ could be translated as either "twice a number" or "two times a number."

STEP 2 What is the meaning of "+ 100"?

The addition sign tells you to add 100 to a number or quantity. This could be translated as either "plus one hundred" or "one hundred more than."

STEP 3 Translate the expression into words.

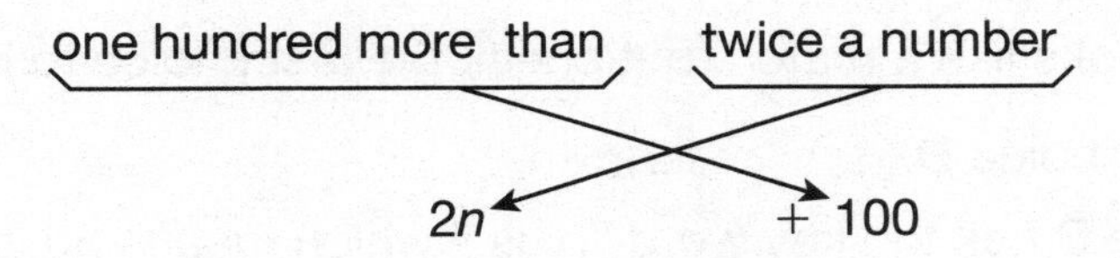

Note: Be as specific as you can be when translating expressions. The expression "twice a number plus one hundred" could mean either $2n + 100$ or $2(n + 100)$. The best translation for $2n + 100$ is "one hundred more than twice a number."

SOLUTION **The algebraic expression $2n + 100$ is the same as the verbal expression "one hundred more than twice a number."**

EXAMPLE 3

Which of the following problem situations matches the expression $12p - 50$?

A. Amy owes a down payment of $50 minus twelve equal payments.

B. Amy owes twelve equal payments times a down payment of $50.

C. Amy owes twelve equal payments reduced by a down payment of $50.

D. Amy owes twelve equal payments plus a down payment of $50.

STRATEGY **Translate each situation into an expression and look for a match to the given expression.**

STEP 1 Translate choice A.

Choice A has the key word "minus," which in this case means that $12p$ is subtracted from 50, or $50 - 12p$.

This does not match the given expression.

STEP 2 Translate choice B.

Choice B has the key word "times," which means to multiply $12p$ by 50, or $50 \cdot 12p$.

This does not match the given expression.

STEP 3 Translate choice C.

Choice C has the key words "reduced by," which means to subtract 50 from $12p$, or $12p - 50$.

This looks like a match, but check the last choice to be sure.

STEP 4 Translate choice D.

Choice D has the key word "plus," which means 50 is added to $12p$, or $12p + 50$.

This does not match the given expression.

SOLUTION **Choice C best matches the expression $12p - 50$.**

COACHED EXAMPLE

Which of the following problem situations matches the expression $3a - 2$?

A. Georgianne is two years more than three times as old as Andrew.

B. Georgianne is three years more than two times as old as Andrew.

C. Georgianne is three years less than two times as old as Andrew.

D. Georgianne is two years less than three times as old as Andrew.

THINKING IT THROUGH

Translate each situation into an expression and compare it with the given expression.

In choice A, the key words are "more than" and "times."

An expression for choice A is ______________________________.

In choice B, the key words are "_____________" and "_____________."

An expression for choice B is ______________________________.

In choice C, the key words are "_____________" and "_____________."

An expression for choice C is ______________________________.

In choice D, the key words are "_____________" and "_____________."

An expression for choice D is ______________________________.

The problem situation that best matches the expression $3a - 2$ is choice _____________.

Lesson Practice

Choose the correct answer.

1. Which verbal expression matches the algebraic expression $5k - 4$?

   A. four more than five times a number
   B. four less than five times a number
   C. four less than a number divided by five
   D. five more than four times a number

2. Which situation matches the expression $36n + 75$?

   A. Amy Lynn owes 36 equal payments minus a down payment of $75.
   B. Amy Lynn owes 36 equal payments to 75 people.
   C. Amy Lynn owes a down payment of $36 and 1 payment of $75.
   D. Amy Lynn owes 36 equal payments plus a down payment of $75.

3. Which verbal expression matches the algebraic expression $7 + \frac{n}{5}$?

   A. seven plus five times a number
   B. seven minus five times a number
   C. seven plus one-fifth of a number
   D. five times a number plus seven

4. Which situation matches the expression $x + 0.07x$?

   A. A computer costs x dollars plus 7% sales tax.
   B. A computer costs x dollars, including 7% sales tax.
   C. A computer costs x dollars less a 7% discount.
   D. A computer costs x dollars in addition to 0.7% sales tax.

5. Which verbal expression matches the algebraic expression $\frac{x}{6} - 4$?

   A. four more than a number divided by six
   B. four less than a number divided by six
   C. four less than six times a number
   D. four more than six times a number

6. Write a verbal expression that matches the algebraic expression $35h + 50$.

   Answer ______________

7. Write a situation that matches the expression $4x + 17$.

   Answer ______________

Lesson

6 Write Inequalities

Getting the Idea

An **inequality** is a statement that compares two quantities. Inequalities may contain any one of these symbols: $>$, $<$, $\geq$, $\leq$, and $\neq$.

Remember:

$>$ means is greater than

$\geq$ means is greater than or equal to

$<$ means is less than

$\leq$ means is less than or equal to

$\neq$ means is not equal to

You can write inequalities to represent real-world situations. Look for key words that can help you translate the situation into an inequality.

Here are some examples of key words.

Problem Situation	Key Words	Expression
The price is more than $50.	more than	$p > 50$
The building's height is less than 125 feet.	less than	$b < 125$
Your height must be at least 40 inches to ride.	at least	$h \geq 40$
Jess will spend no more than $12 on lunch.	no more than	$l \leq 12$
The company's sales are not equal to its expenses.	not equal to	$s \neq e$

EXAMPLE 1

A game involves two rolls of a number cube with faces numbered 1 through 6. A player wins if the sum of the two rolls is less than 6. On his first roll Rodney gets a 2. Write an inequality that can be used to find the numbers Rodney needs to get on the second roll in order to win.

STRATEGY **Choose and identify a variable and look for key words.**

STEP 1 Choose a variable.

Let x represent the range of numbers Rodney needs on his second roll. (Any variable would work.)

STEP 2 What are the key words and what do they represent?

The word "sum" indicates addition.

The words "less than" indicate that the inequality involves the $<$ symbol.

STEP 3 Translate the words into an inequality.

"The sum of the rolls" can be translated as $2 + x$.

"Must be less than 6" can be translated as < 6.

SOLUTION **The inequality $2 + x < 6$, or $x + 2 < 6$, can be used to find the numbers Rodney needs to get on his second roll.**

EXAMPLE 2

Michele wants to buy 3 DVDs and a CD without spending more than \$80. Each DVD costs d dollars and the CD costs \$14. Write an inequality to represent the situation.

STRATEGY **Use the variable and look for key words.**

STEP 1 Write what you know.

The price is the same on each DVD.

The CD costs \$14.

Michele does not want to spend more than \$80.

STEP 2 Write what goes on each side of the inequality.

$3d + 14$ goes on one side of the inequality.

80 goes on the other side of the inequality.

STEP 3 Choose the correct inequality symbol.

Not wanting to spend more than \$80 means that Michele can spend up to and including \$80, so use $\leq$.

SOLUTION **The inequality $3d + 14 \leq 80$ represents the situation.**

COACHED EXAMPLE

Nadia has saved $100 to spend on vacation. She would like to have a total of at least $150 to spend on vacation. Let *s* represent the additional amount of money she needs to save. Write an inequality to represent the situation.

THINKING IT THROUGH

The key word "additional" indicates that you should ____________ *s* and 100.

An expression for the total amount Nadia needs to save is ____________.

The key words "at least" mean that the inequality symbol is ____________.

An inequality that represents the situation is ____________.

Lesson Practice

Choose the correct answer.

1. The inequality $a \geq 65$ represents the age in years, a, at which ABC Company employees are eligible for retirement. Which statement is true about this situation?

   A. Employees are eligible to retire when they are less than 65 years old.

   B. Employees must retire when they turn 65 years old.

   C. Employees are eligible to retire when they are 65 years old or older.

   D. Employees are eligible to retire when they are older than 65 years old but not before.

2. A taxi charges \$4 plus \$3 for each mile traveled. Melissa does not want to spend more than \$15 on taxi fare. Let m represent the number of miles in a trip. Which of the following inequalities represents this situation?

   A. $4 + 3m < 15$

   B. $4 + 3m \leq 15$

   C. $4 + 3m \geq 15$

   D. $4 + 3m \neq 15$

3. Yolanda has \$38 in a bank account. She wants to make two equal deposits, after which she wants her account to have a balance of at least \$100. Which inequality represents the number of dollars, d, Yolanda could deposit each time?

   A. $d + 38 \leq 100$

   B. $d + 38 \geq 100$

   C. $2d + 38 \leq 100$

   D. $2d + 38 \geq 100$

4. Jane is buying bagels and cream cheese for a family brunch. At most, she can spend \$25. She will spend a total of \$4 on cream cheese. Each bagel costs \$0.90. Which inequality could be used to determine b, the number of bagels she could buy?

   A. $0.9b + 4 \leq 25$

   B. $0.9b - 4 \leq 25$

   C. $4 - 0.9b \leq 25$

   D. $0.9b + 4 \neq 25$

5. Write a statement that can be represented by the inequality $s \leq 35$.

   Answer ______________________________

   ____

6. A theater has a capacity of less than 750 seats. So far, 428 tickets have been sold for next week's concert. Let n represent the number of tickets that remain to be sold. Write an inequality to represent this situation.

   Answer ____________________

Lesson

7 Solve and Graph Linear Inequalities

8.A.13, 8.A.14, 8.G.19

Getting the Idea

To solve an inequality, follow the same steps as in solving an equation. What you do to one side of the inequality you must do to the other side. The solution will be a set of numbers instead of one number. This set of numbers is called a **solution set**.

You can graph the solution set of an inequality on a number line. The value shown in the inequality is drawn with a circle.

If the value is part of the solution ($\leq$ or $\geq$), fill in the circle.

If the value is not part of the solution ($<$ or $>$), use an open circle.

Draw an arrow to represent the solution set.

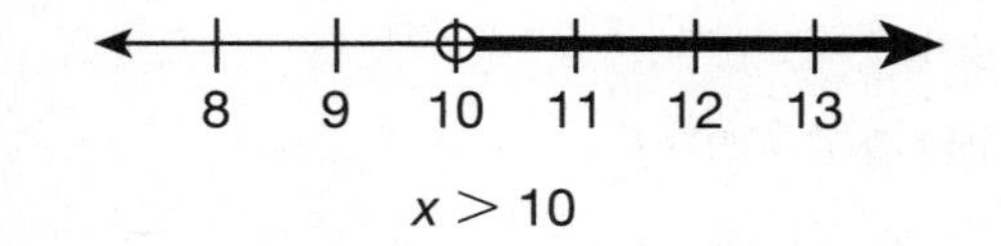

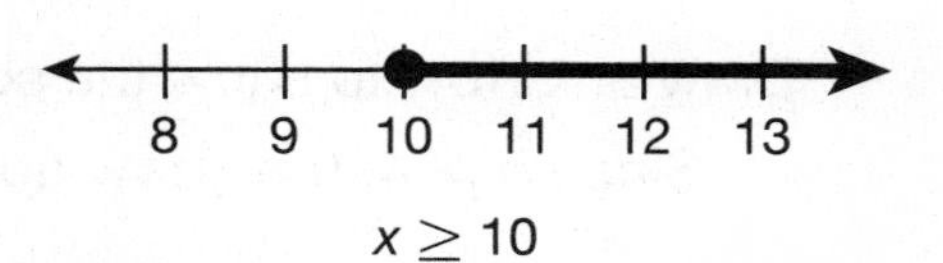

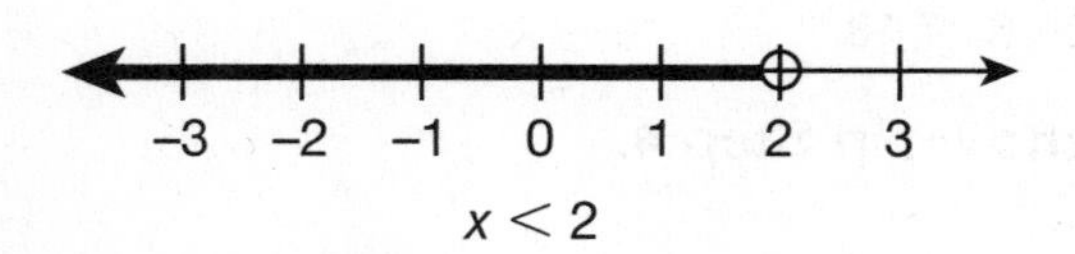

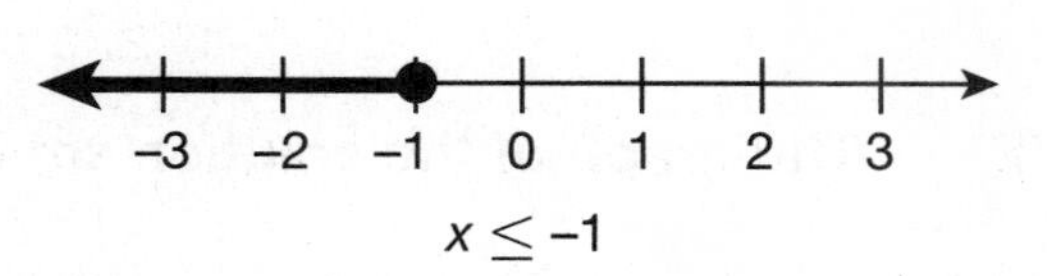

EXAMPLE 1

Graph the solution set of $4b + 2 > 18$.

STRATEGY **Use inverse operations to isolate the variable. Then graph the solution.**

STEP 1 Subtract 2 from both sides of the inequality.

$$4b + 2 - 2 > 18 - 2$$

$$4b > 16$$

STEP 2 Divide both sides of the inequality by 4 to isolate the variable.

$$\frac{4b}{4} > \frac{16}{4}$$

$$b > 4$$

STEP 3 Locate 4 on a number line.

Draw a number line that has 4 in the middle. Since 4 is not part of the solution, draw an open circle.

1 2 3 4 5 6 7 8

STEP 4 Draw an arrow to show the solution set.

Since $b > 4$, the arrow goes to the right from 4.

1 2 3 4 5 6 7 8

SOLUTION **The graph of the solution set is shown in Step 4.**

One difference between solving inequalities and equations is that when you multiply or divide both sides of an inequality by a negative number, you must reverse the direction of the inequality symbol. Consider the inequality $5 > 3$.

If you multiply both sides of the inequality by -2, the result is -2×5, or -10, on the left side of the inequality, and -2×3, or -6, on the right side. The inequality $-10 > -6$ is not true. To have a true inequality, reverse the direction of the inequality symbol: $-10 < -6$.

Now consider this inequality: $14 \leq 35$.

If you divide both sides by -7, the result is -2 on the left side and -5 on the right side. The inequality $-2 \leq -5$ is not true. In order to have a true inequality, reverse the direction of the inequality: $-2 \geq -5$.

Remember, when you multiply or divide both sides of an inequality by a negative number, you reverse the inequality.

EXAMPLE 2

Solve $2(4 - x) \leq 14$ and graph the solution.

STRATEGY **Use inverse operations to isolate the variable.**

STEP 1 Use the distributive property.

$$2(4 - x) \leq 14$$
$$8 - 2x \leq 14$$

STEP 2 Subtract 8 from both sides of the inequality.

$$8 - 2x \leq 14$$
$$8 - 2x - 8 \leq 14 - 8$$
$$-2x \leq 6$$

STEP 3 Divide both sides by -2. Reverse the direction of the inequality.

$$-2x \leq 6$$
$$\frac{-2x}{-2} \geq \frac{6}{-2}$$
$$x \geq -3$$

STEP 4 Graph the inequality.

Locate -3 on a number line.

Use a closed circle because -3 is part of the solution.

Draw an arrow to the right to indicate all numbers greater than or equal to -3.

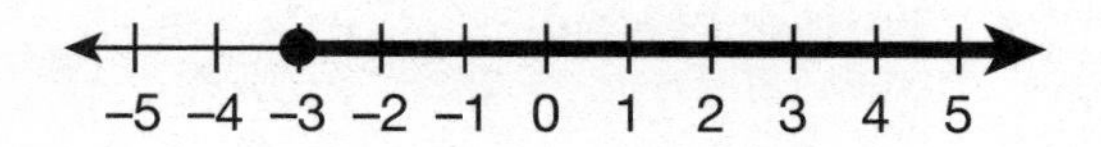

SOLUTION **The solution of $2(4 - x) \leq 14$ is $x \geq -3$. Its graph is shown in Step 4.**

COACHED EXAMPLE

Solve: $x + 8 \leq 5 - 2x$

THINKING IT THROUGH

Both sides of the inequality have variable terms.

Add $2x$ to both sides to eliminate $-2x$ from the right side of the inequality.

____

Begin to isolate the variable by subtracting ____________ from both sides.

____

Complete isolating the variable by dividing both sides by ____________.

The solution is ___________.

Lesson Practice

Choose the correct answer.

1. Which is one of the solutions of this inequality?

$$4r + 8 \leq 24$$

A. 3
B. 6
C. 9
D. 12

2. What is the solution set of this inequality?

$$2(z - 3) \geq 14$$

A. $z \geq 4$
B. $z \leq 4$
C. $z \geq 10$
D. $z \leq 10$

3. What is the solution set of this inequality?

$$8 - 4r \leq 32$$

A. $r \geq -6$
B. $r > -6$
C. $r \leq -6$
D. $r < 6$

4. What is the solution set of this inequality?

$$7x + 2 > 5x + 6$$

A. $x > -2$
B. $x < -2$
C. $x > 2$
D. $x < 2$

5. Which graph represents the solution set of this inequality?

$$2(x + 2) > 10$$

A. 0 1 2 3 4 5 6 7 8 9 10
B. 0 1 2 3 4 5 6 7 8 9 10
C. 0 1 2 3 4 5 6 7 8 9 10
D. 0 1 2 3 4 5 6 7 8 9 10

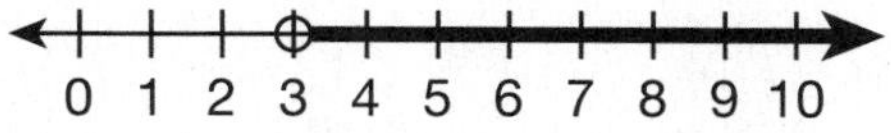

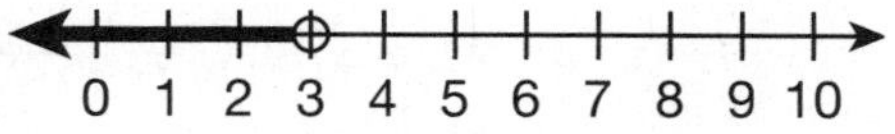

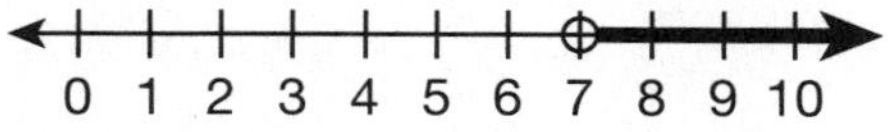

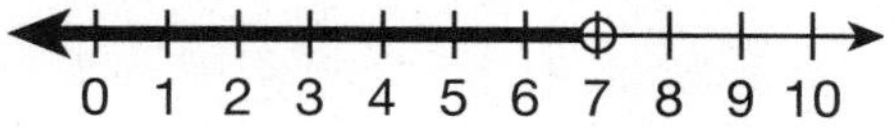

6. Which graph represents the solution to this inequality?

$$4x + 1 > x - 5$$

A. –5 –4 –3 –2 –1 0 1 2 3 4 5
B. –5 –4 –3 –2 –1 0 1 2 3 4 5
C. –5 –4 –3 –2 –1 0 1 2 3 4 5
D. –5 –4 –3 –2 –1 0 1 2 3 4 5

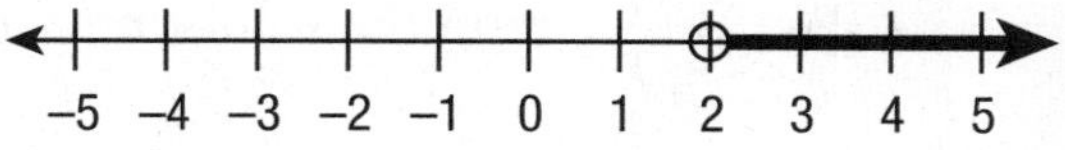

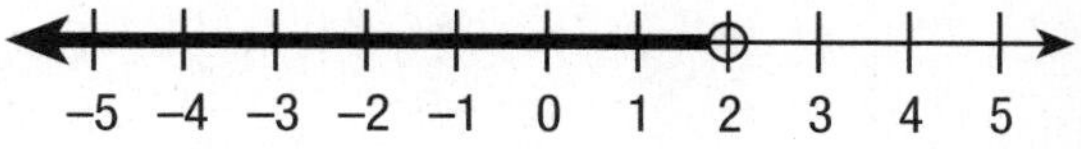

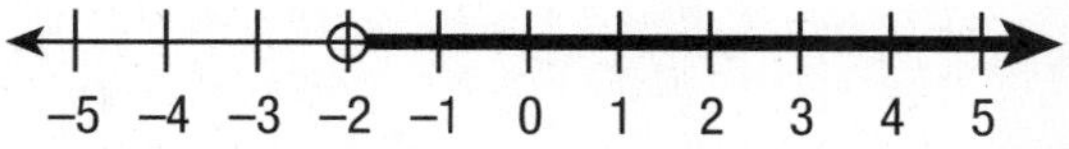

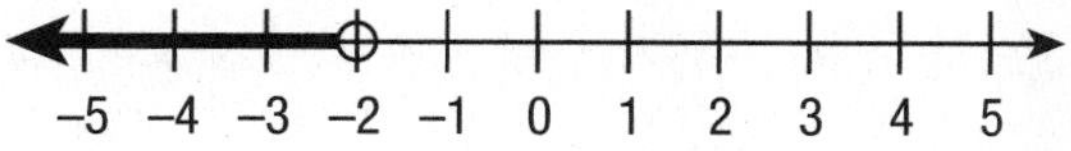

7. Solve: $4 - 2x < 6$

Answer ________________

8. Solve: $3(x - 3) > 9$

Answer ________________

EXTENDED-RESPONSE QUESTION

9. Consider this inequality.

$x \leq -2x + 12$

Part A What is its solution set?

Answer ________________

Part B Graph the solution on the number line below.

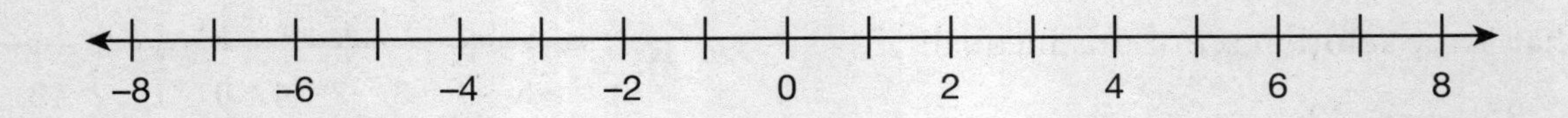

8 Graph Situations

8.A.3, 8.A.4

Getting the Idea

A **graph** shows a relationship between sets of data. It is important to be able to interpret the information given in graphs.

The graph below shows the relationship between time and the distance covered by Luz during a jog.

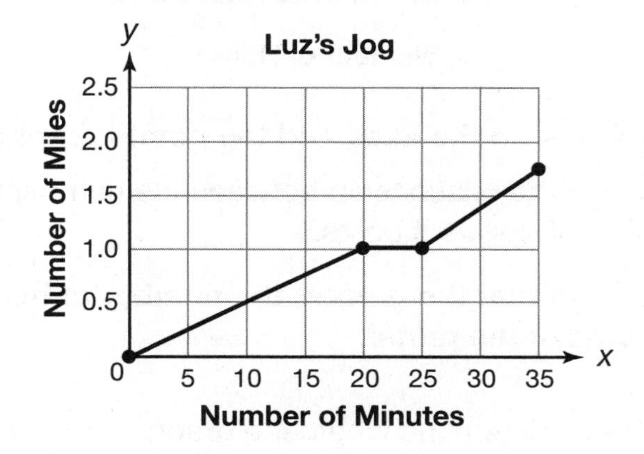

Here is some information given in the graph.

- For the first 20 minutes, she jogged a distance of 1 mile, so she jogged an average rate of 1 mi $\div$ 20 min = 1 mi $\div \frac{1}{3}$ h, or 3 miles per hour.
- For the next 5 minutes, she did not cover any additional distance. This is indicated by the horizontal line segment from 20 min to 25 min. This might mean that she stopped jogging for 5 minutes or that she jogged in place for 5 minutes.
- For the next 10 minutes, she jogged $\frac{3}{4}$ mile, so she jogged at an average rate of 0.75 mi $\div$ 10 min = 0.75 mi $\div \frac{1}{6}$ h, or 4.5 miles per hour.

EXAMPLE 1

What situation is represented by this graph?

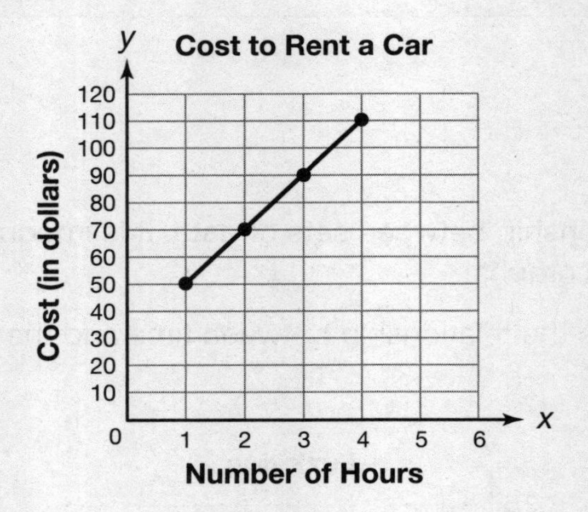

STRATEGY **Look at the labels on the axes and the points graphed.**

The graph shows the relationship between the number of hours a car is rented and the number of dollars it costs.

SOLUTION **The graph shows that the greater the number of hours a car is rented, the greater the cost of the rental.**

It is also important to be able to determine whether a graph accurately represents a given situation.

EXAMPLE 2

Sylvia made 20 gallons of mint tea to sell at the town fair. Within the first 2 hours she had sold almost all of her tea. Which graph below best represents this situation?

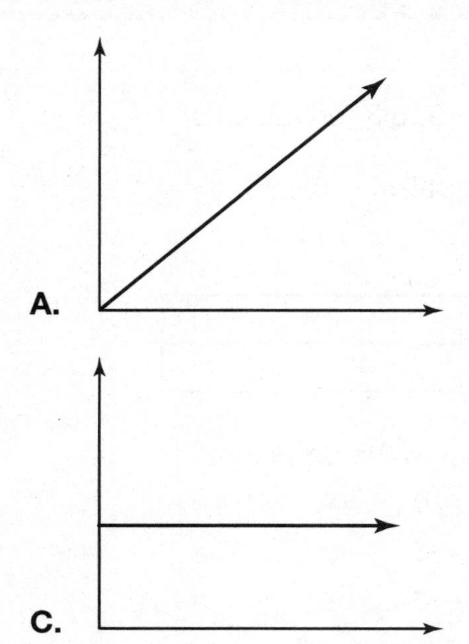

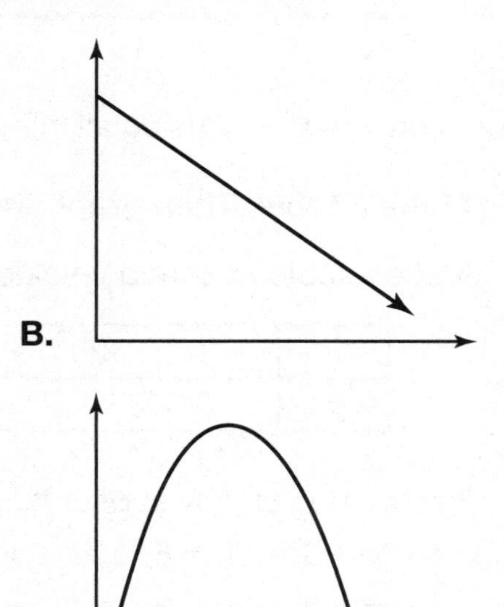

STRATEGY **Match the graph to the given situation.**

STEP 1 Interpret the meaning of the given situation.

As Sylvia gets more customers, what will happen to the amount of tea she has available for sale? The more customers Sylvia has, the less tea she will have available for sale. A graph of the situation should show a decreasing line.

STEP 2 Find the graph that matches your interpretation.

Start by imagining that the horizontal axis of each graph represents Customers, and the vertical axis represents Tea Available for Sale. Now analyze each graph.

Graph A: The line in Graph A is rising, which is the opposite of the situation.

Graph B: The line in Graph B is falling. This matches the situation, but you should check the remaining choices to be sure.

Graph C: The line in Graph C is constant, which does not match the situation.

Graph D: The line in Graph D rises and then falls, which does not match the situation.

Graph B is the only graph that could represent how much tea Sylvia has available for sale at any given time.

SOLUTION **Graph B best represents the situation.**

A relationship may sometimes be given to you in the form of an equation. To graph that relationship, you can use the equation to create **ordered pairs** of the form (x, y), where x is the horizontal coordinate of the point and y is the vertical coordinate.

EXAMPLE 3

Graph the equation $y = x^2 - 1$ using all integers from -3 to 3 for values of x.

STRATEGY **Make a table. Then graph the ordered pairs.**

STEP 1 Make a table of x- and y-values.

x	-3	-2	-1	0	1	2	3
y							

STEP 2 Find values of y by substituting x-values into the equation.

$y = -3^2 - 1 = 8$ $\quad$ $y = 1^2 - 1 = 0$

$y = -2^2 - 1 = 3$ $\quad$ $y = 2^2 - 1 = 3$

$y = -1^2 - 1 = 0$ $\quad$ $y = 3^2 - 1 = 8$

$y = 0^2 - 1 = -1$

STEP 3 Fill in the y-values in the table.

x	-3	-2	-1	0	1	2	3
y	8	3	0	-1	0	3	8

STEP 4 Plot the points and connect them.

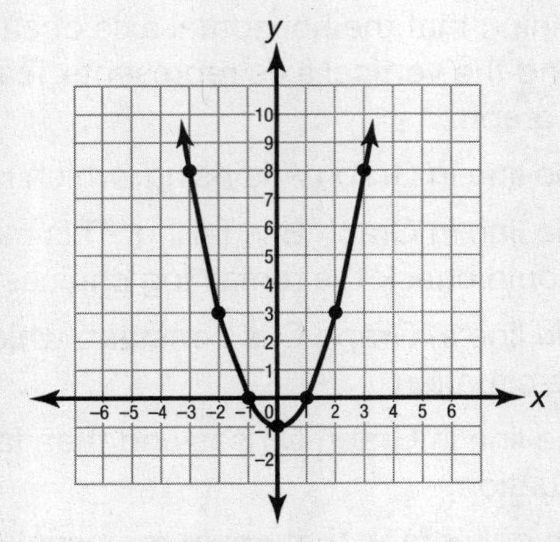

SOLUTION **The graph of the equation is shown in Step 4.**

COACHED EXAMPLE

This graph shows the relationship between time and the amount of water in a tank that is being drained for cleaning.

How is the amount of water in the tank changing over time?

After how many minutes will there be 5 gallons left in the tank?

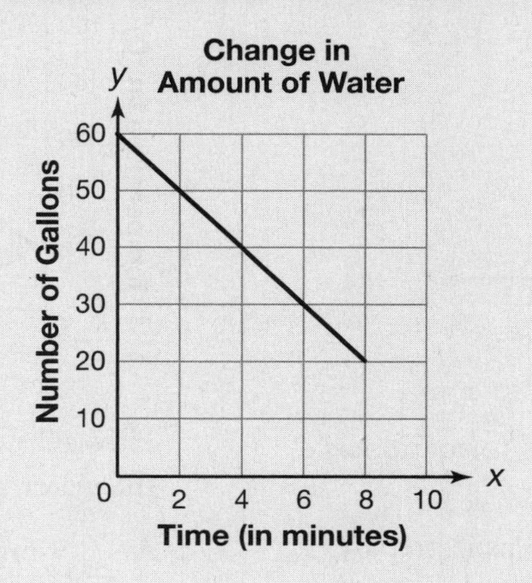

THINKING IT THROUGH

The line is falling from left to right, so the amount of water in the tank is ____________ over time.

Each point on the segment is an ordered pair of the form (Minute, Gallons). Extend the line to the point that has a Gallons coordinate of ____________.

There will be 5 gallons of water left in the tank after ____________ minutes.

There will be 5 gallons of water left in the tank after ____________ minutes.

Lesson Practice

Choose the correct answer.

1. What situation is represented by this graph?

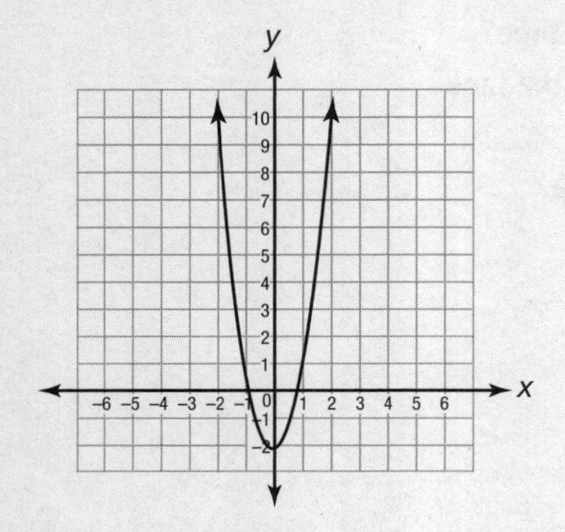

A. A flock of geese changes formation as it travels north.

B. An airplane slows down as it prepares to land.

C. An eagle swoops down on its prey and returns to its nest.

D. A boulder gains speed as it rolls down a mountain.

2. This graph shows the change in water temperature of a pot of tap water on a stove-top burner.

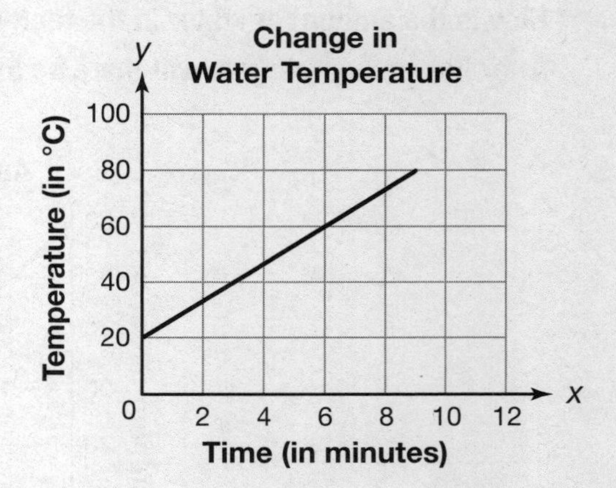

How many total minutes will it take for the water to reach 100°C?

A. 9 minutes

B. 10 minutes

C. 11 minutes

D. 12 minutes

3. Which graph represents $y = 2x - 3$?

A.

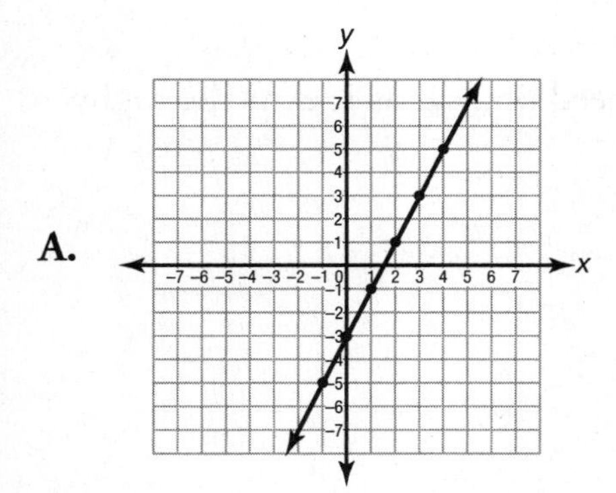

B.

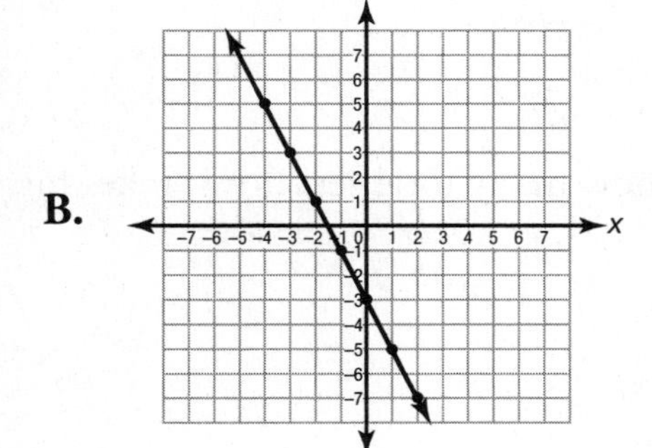

C.

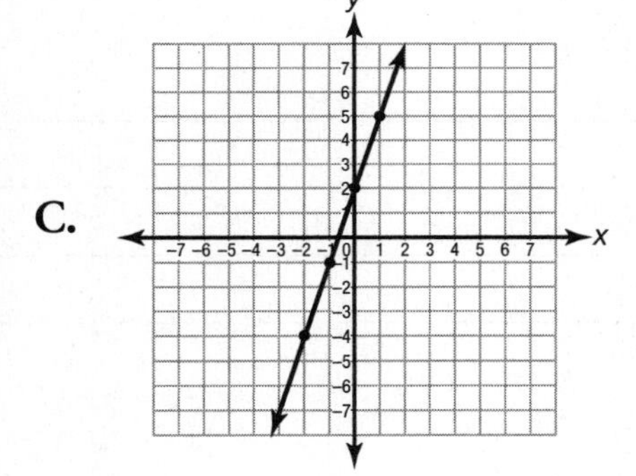

D.

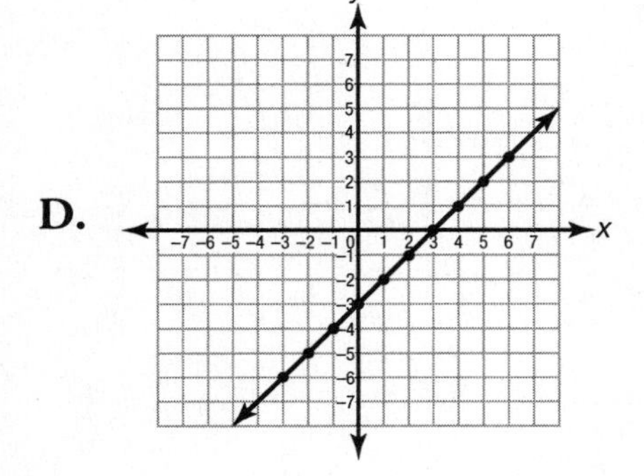

4. Which equation has this graph?

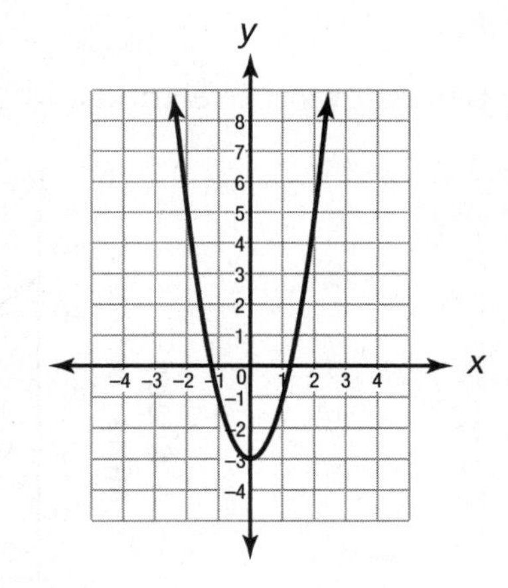

A. $y = x^2 - 3$

B. $y = 2x^2 - 3$

C. $y = 3x^2 - 2$

D. $y = 2x - 3$

5. Which is the equation for this table?

x	−3	−2	−1	0	1
y	5	3	1	−1	−3

A. $y = 3x + 4$

B. $y = 2x^2$

C. $y = -2x - 1$

D. $y = -\frac{1}{2}x$

EXTENDED-RESPONSE QUESTION

6. This graph shows the relationship between time and the speed of a car as it exits the highway.

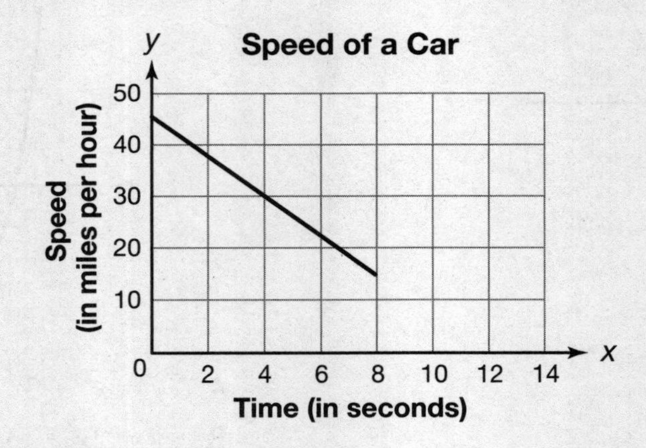

Part A Use the information in the graph to determine how many total seconds it takes for the car to come to a complete stop.

Part B Explain how you determined your answer.

____

____

____

____

Lesson

9 Add and Subtract Polynomials

8.A.5, 8.A.7

Getting the Idea

A **polynomial** is an expression consisting of either one term, or the sum or difference of two or more terms. A **term** is a number, a variable, or the product of a number and one or more variables. Here are some examples of terms: 3, n, $5x$, $-2a$, $6xy$, and $4n^2$.

Terms that have the same variable(s) raised to the same power(s) are called **like terms**. For example, $2x$ and $3x$ are like terms; $9y^2$ and $-7y^2$ are like terms; but x^2y and xy^2 are not like terms.

Some polynomials have special names:

- A **monomial** has only one term. For example, -7, $3c$, $8x^2$, and $5jk$ are monomials.
- A **binomial** is the sum or difference of two unlike terms. For example, $x + 1$ and $n^2 - 9$ are binomials.
- A **trinomial** is the sum and/or difference of three unlike terms. For example, $n^2 - 7n + 12$ and $3x^2 + 10x + 8$ are trinomials.

Note: For an expression to be a polynomial, each of its terms must have a positive exponent.

You can use algebra tiles to model addition and subtraction of polynomials. Algebra tiles are pictured below, with their values shown inside. The algebra tiles you use in class may have one color to indicate positive tiles and another color to indicate negative tiles.

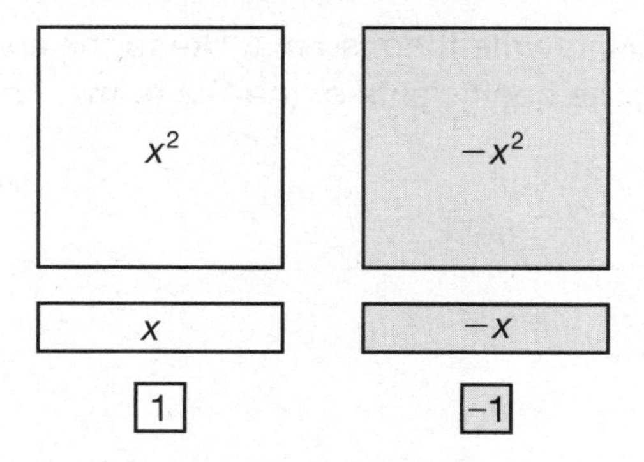

When two tiles have the same shape but different signs, they form a zero pair. That is, their sum is zero and they can be removed from the collection of tiles.

EXAMPLE 1

Use algebra tiles to model $(2x + 3) + (x^2 - 2)$.

STRATEGY **Model each binomial addend. Arrange the addends vertically, and then add.**

STEP 1 Model each binomial.

$2x + 3 =$ x x 1 1 1

$x^2 - 2 =$ x^2 −1 −1

STEP 2 Group tiles with the same shape and remove any zero pairs.

x^2 x x 1 1 1 −1 −1

SOLUTION $(2x + 3) + (x^2 - 2) = x^2 + 2x + 1$

To add or subtract polynomials, rewrite them so that like terms are grouped together. Then combine like terms by adding the coefficients of the like terms. For example:

$7x + 4x = (7 + 4)x = 11x$

$-5m^2 + 2m^2 = (-5 + 2)m^2 = -3m^2$

EXAMPLE 2

Add: $(a^2 - 6a + 1) + (3a^2 - 5a - 6)$

STRATEGY **Rewrite the expression so that like terms are grouped together. Then combine like terms.**

STEP 1 Rewrite the expression so that like terms are grouped together.

$a^2 + 3a^2 - 6a - 5a + 1 - 6$

STEP 2 Combine like terms.

$$\underbrace{a^2 + 3a^2}_{4a^2}\ \underbrace{-\ 6a - 5a}_{-\ 11a}\ \underbrace{+\ 1 - 6}_{-\ 5}$$

SOLUTION $\mathbf{(a^2 - 6a + 1) + (3a^2 - 5a - 6) = 4a^2 - 11a - 5}$

To subtract polynomials, add the opposite of the second polynomial. To find the opposite of a polynomial, change the sign of each term. This is the same as multiplying each term by -1.

$-(-2n^2 - 3n + 6) = 2n^2 + 3n - 6$

$-(6x^2 + 4x - 3) = -6x^2 - 4x + 3$

EXAMPLE 3

Subtract: $(4a^2 - 3a - 6) - (2a^2 - 6a - 5)$

STRATEGY **Find the opposite of $2a^2 - 6a - 5$ and add to $4a^2 - 3a - 6$.**

STEP 1 Find the opposite of $2a^2 - 6a - 5$.

$-(2a^2 - 6a - 5) = -2a^2 + 6a + 5$

So, $(4a^2 - 3a - 6) - (2a^2 - 6a - 5)$ can be written as:

$4a^2 - 3a - 6 - 2a^2 + 6a + 5$

STEP 2 Rewrite the expression so that like terms are grouped together.

$4a^2 - 2a^2 - 3a + 6a - 6 + 5$

STEP 3 Combine like terms.

$$\underbrace{4a^2 - 2a^2}_{2a^2}\ \underbrace{-\ 3a + 6a}_{+\ 3a}\ \underbrace{-\ 6 + 5}_{-\ 1}$$

SOLUTION $\mathbf{(4a^2 - 3a - 6) - (2a^2 - 6a - 5) = 2a^2 + 3a - 1}$

COACHED EXAMPLE

Find the perimeter of the triangle shown below.

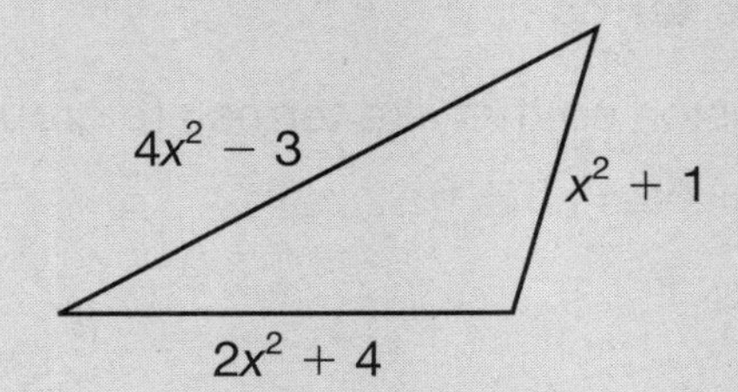

THINKING IT THROUGH

Recall that to find the perimeter of a polygon, you need to add the lengths of all its sides.

P = ____________ + ____________ + ____________

Next, rewrite the expression so that like terms are grouped together.

P = (_________ + _________ + _________) + (_________ + _________ + _________)

Now combine like terms to find an expression that represents the perimeter.

P = ____________ + ____________

The triangle has a perimeter of ____________ units.

Lesson Practice

Choose the correct answer.

1. What is the sum of the polynomials modeled below?

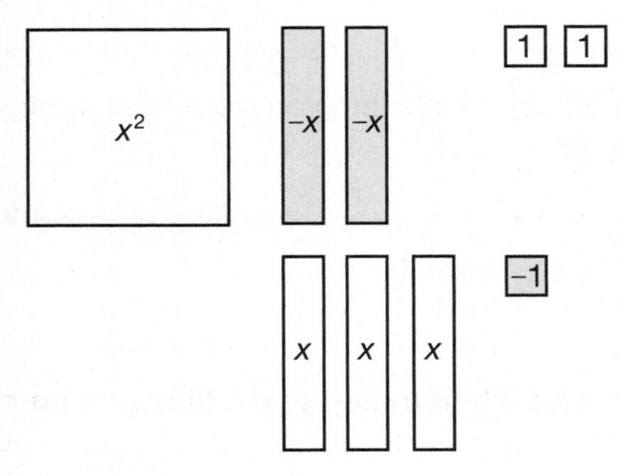

A. $x^2 + 5x + 3$
B. $x^2 - 2x + 1$
C. $x^2 - x - 1$
D. $x^2 + x + 1$

2. Find the sum.

$(c^2 - 6c) + (7c + 5)$

A. $4c^2 + 2c$
B. $c^2 + c + 5$
C. $c^2 - c + 5$
D. $c^2 + 7c + 5$

3. Find the difference.

$(9p - 15) - (2p + 3)$

A. $7p - 12$
B. $7p - 18$
C. $11p - 12$
D. $11p - 18$

4. Simplify: $(3x^2 + 3x + 1) + (7x^2 + 2x + 5)$

A. $10x^4 + 5x^2 + 6$
B. $10x^2 + 5x + 5$
C. $10x^2 + 5x + 6$
D. $21x^2 + 6x + 5$

5. Simplify: $(11a^2 - 5a + 6) - (3a^2 - 2a + 4)$

A. $-8a^2 - 3a + 2$
B. $8a^2 - 3a + 2$
C. $8a^2 - 7a + 10$
D. $14a^2 - 3a + 2$

6. What is an expression for the perimeter of this rectangle?

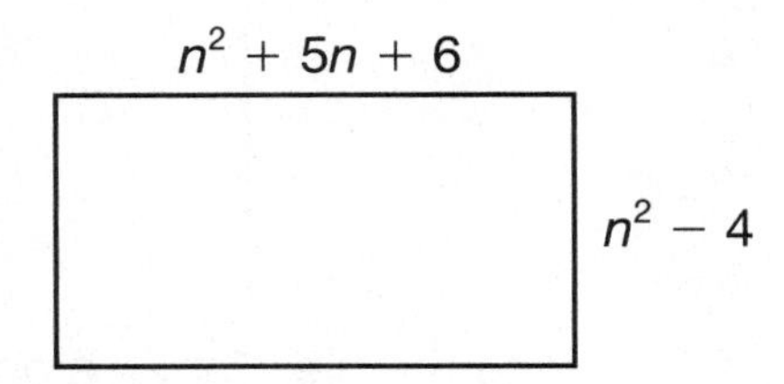

Answer ______________

7. Each side of an equilateral triangle measures $3x^2 - 5$ inches. What is the perimeter of this triangle?

Answer ______________

10 Multiply and Divide Monomials

8.N.1, 8.A.5, 8.A.6, 8.A.9

Getting the Idea

You can use algebra tiles to model multiplication of monomials, but you must remember the exponent rule for multiplying powers with the same base.

EXAMPLE 1

Use algebra tiles to find the product of $2x$ and $3x$.

STRATEGY **Place the tiles along the edges of a frame. Then multiply two tiles at a time.**

STEP 1 Make a frame.

STEP 2 Place the tiles for $2x$ along the vertical edge of the frame and the tiles for $3x$ along the horizontal edge.

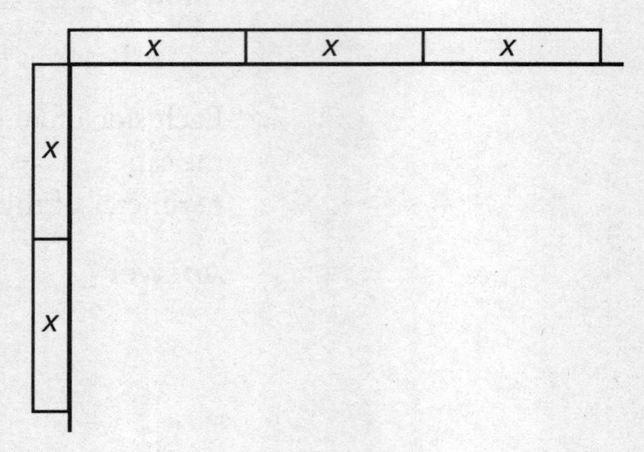

STEP 3 Multiply the first x along the vertical edge and the first x along the horizontal edge.

Remember that $x = x^1$.

To multiply two powers with the same base, add the exponents.

$x \cdot x = x^1 \cdot x^1 = x^{1+1} = x^2$

Place x^2 inside the frame.

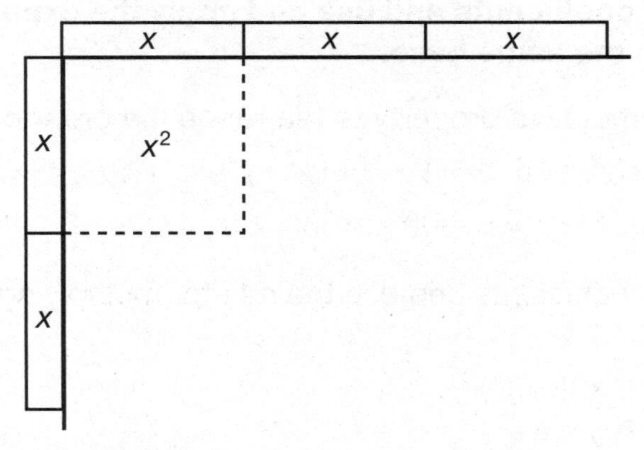

STEP 4 Complete the multiplication.

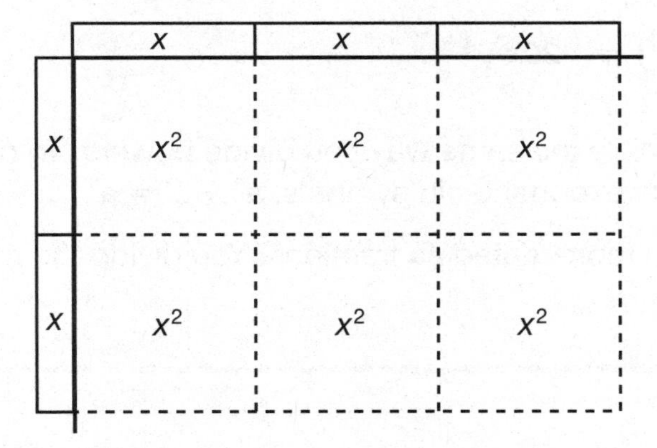

STEP 5 Add the terms.

There are 6 x^2 tiles, or $6x^2$.

SOLUTION **The product of $2x$ and $3x$ is $6x^2$.**

You can multiply monomials the same way you multiply powers. To multiply two powers with the same base, add the exponents. In symbols, $a^s \cdot a^t = a^{s+t}$.

EXAMPLE 2

Multiply: $(5x^2y)(-4x^3y^2)$

STRATEGY **Multiply the coefficents and use and apply the exponent rule for multiplying powers with the same base.**

STEP 1 Use the commutative property to rearrange the order of the factors.

$$(5x^2y)(-4x^3y^2) = 5 \cdot x^2 \cdot y \cdot -4 \cdot x^3 \cdot y^2$$
$$= (5 \cdot -4)(x^2 \cdot x^3)(y \cdot y^2)$$

STEP 2 Multiply the coefficients and use the rule for multiplying powers with the same base.

$$(5 \cdot -4)(x^2 \cdot x^3)(y \cdot y^2)$$
$$(-20)(x^{2+3})(y^{1+2})$$
$$(-20)(x^5)(y^3)$$

SOLUTION $\mathbf{(5x^2y)(-4x^3y^2) = -20x^5y^3}$

You can also divide monomials the same way you divide powers. To divide two powers with the same base, subtract the exponents. In symbols, $a^s \div a^t = a^{s-t}$.

Division problems are often represented as fractions. You divide the numerator by the denominator.

EXAMPLE 3

Divide: $\frac{36n^8}{9n^5}$

STRATEGY **Divide the coefficients and apply the exponent rule for dividing powers with the same base.**

STEP 1 Rewrite the fraction to see the division of the coefficients and the division of the variables.

$$\frac{36n^8}{9n^5} = \frac{36 \cdot n^8}{9 \cdot n^5}$$
$$= \frac{36}{9} \cdot \frac{n^8}{n^5}$$

STEP 2 Divide.

$$\frac{36}{9} \cdot \frac{n^8}{n^5} = 4 \cdot n^{8-5}$$
$$= 4 \cdot n^3$$

SOLUTION $\frac{36n^8}{9n^5} = 4n^3$

When you divide a polynomial by a monomial, you divide each term in the polynomial by the monomial.

When dividing a polynomial by a monomial, you may sometimes end up with a variable raised to the zero power. For example, $x^7 \div x^7 = x^{7-7} = x^0$. You know that any nonzero number divided by itself is 1, so $x^7 \div x^7$ is also equal to 1. Since $x^7 \div x^7$ equals both x^0 and 1, then $x^0 = 1$. In general, any nonzero number with an exponent of zero is equal to 1.

EXAMPLE 4

Divide: $\frac{15x^3 - 5x^2 + 10x}{5x}$

STRATEGY **Divide each term in the numerator by the denominator.**

STEP 1 Rewrite the fraction to see each division.

$$\frac{15x^3 - 5x^2 + 10x}{5x} = \frac{15x^3}{5x} - \frac{5x^2}{5x} + \frac{10x}{5x}$$

STEP 2 Divide.

$$\frac{15x^3}{5x} - \frac{5x^2}{5x} + \frac{10x}{5x}$$
$$3x^{3-1} - x^{2-1} + 2x^{1-1}$$
$$3x^2 - x^1 + 2x^0$$
$$3x^2 - x + 2$$

SOLUTION $\frac{15x^3 - 5x^2 + 10x}{5x} = 3x^2 - x + 2$

COACHED EXAMPLE

Divide: $\frac{12x^5 + 18x^4 - 6x^3}{-3x}$

THINKING IT THROUGH

Divide each term of the polynomial in the numerator by the monomial in the denominator.

Rewrite the expression as three fractions to see each division.

____________ + ____________ − ____________

Perform the first division.

$12 \div -3 =$ ____________ and $x^5 \div x^1 =$ ____________, so $\frac{12x^5}{-3x} =$ ____________.

Perform the second division.

$18 \div -3 =$ ____________ and $x^4 \div x =$ ____________, so $\frac{18x^4}{-3x} =$ ____________.

Perform the third division.

$6 \div -3 =$ ____________ and $x^3 \div x =$ ____________, so $\frac{6x^3}{-3x} =$ ____________.

Write the results of the three divisions.

____________ + ____________ − ____________

In the second term you are adding a negative, so change the operation sign from addition to ____________ and drop the negative sign.

In the third term you are subtracting a negative, so change the operation sign from subtraction to ____________ and drop the negative sign.

$\frac{12x^5 + 18x^4 - 6x^3}{-3x} =$ ____________

Lesson Practice

Choose the correct answer.

1. Find the product.

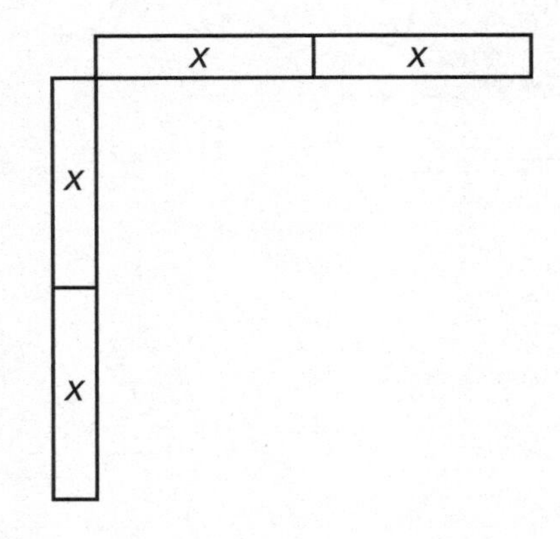

A. 4
B. $4x$
C. $4x^2$
D. x^4

2. What is the product of $9n^3$ and $6n^3$?

A. $15n^6$
B. $54n^3$
C. $54n^6$
D. $54n^9$

3. Simplify: $(-3a^6)(-4a)$

A. $-12a^6$
B. $-12a^7$
C. $12a^6$
D. $12a^7$

4. Simplify: $(7x^3y)(-4x^2y^4)$

A. $-28x^6y^4$
B. $-28x^5y^5$
C. $28x^6y^4$
D. $28x^5y^5$

5. What is $24x^3$ divided by $8x$?

A. $3x^2$
B. $3x^3$
C. $16x^2$
D. $16x^3$

6. Simplify: $\frac{-32n^4}{-4n^3}$

A. $-8n$
B. $-28n$
C. $8n$
D. $8n^7$

7. Simplify: $\frac{14x^3 + 35x^2 + 21x}{7}$

A. $2x^2 + 5x + 1$
B. $2x^2 + 5x + 3$
C. $2x^3 + 5x^2 + 3x$
D. $2x^4 + 5x^3 + 3x^2$

8. Simplify: $\dfrac{25n^5 - 15n^3 + 30n}{-5n}$

A. $-5n^4 - 3n^2 - 6$

B. $-5n^4 + 3n^2 - 6$

C. $-5n^4 + 3n^2 + 6$

D. $5n^4 + 3n^2 + 6$

9. Simplify: $(8a^2b)(-3ab^2)$

Answer ______________

10. Simplify: $\dfrac{27x^3y^4}{-3xy^3}$

Answer ______________

EXTENDED-RESPONSE QUESTION

11. Yukio simplified the expression $(6x^3y^2)(3x^2y)$. Her answer was $2xy$.

Part A Is Yukio's answer correct?

____

Part B Explain your answer. If you think she is incorrect, provide the correct simplification.

Lesson

11 Multiply Binomials

8.A.8

Getting the Idea

Before you multiply binomials, you should understand how to multiply a binomial by a monomial. To multiply a binomial by a monomial, apply the **distributive property**. The distributive property states that when you multiply the sum of two terms by a factor, you can multiply each term by that factor and add the products:

$a(b + c) = ab + ac$

The distributive property also applies to subtraction:

$a(b - c) = ab - ac$

EXAMPLE 1

Multiply: $5x(3x - 2)$

STRATEGY **Apply the distributive property.**

STEP 1 Apply the distributive property.

$5x(3x - 2) = 5x(3x) - 5x(2)$

STEP 2 Simplify.

$5x(3x) - 5x(2) = 15x^2 - 10x$

SOLUTION $\mathbf{5x(3x - 2) = 15x^2 - 10x}$

To multiply two binomials, you multiply each term of the first binomial by each term of the second binomial. In other words, the distributive property states that:

$(a + b)(c + d) = a(c + d) + b(c + d)$

EXAMPLE 2

Multiply: $(2x - 5)(x + 4)$

STRATEGY **Use the distributive property to multiply the two binomials.**

STEP 1 Multiply $2x$, the first term in $(2x - 5)$, by each term in $(x + 4)$.

$2x \cdot x = 2x^2$

$2x \cdot 4 = 8x$

STEP 2 Multiply -5, the second term in $(2x - 5)$, by each term in $(x + 4)$. Remember to include the negative sign.

$-5 \cdot x = -5x$

$-5 \cdot 4 = -20$

STEP 3 Add the products together, combining like terms.

$2x^2 + 8x - 5x - 20 = 2x^2 + 3x - 20$

SOLUTION $\mathbf{(2x - 5)(x + 4) = 2x^2 + 3x - 20}$

You can check your work in Example 2 by using the **FOIL method**, a commonly used process for multiplying two binomials. FOIL stands for First, Outer, Inner, Last. This graphic shows the first, outer, inner, and last terms in the binomials in Example 2.

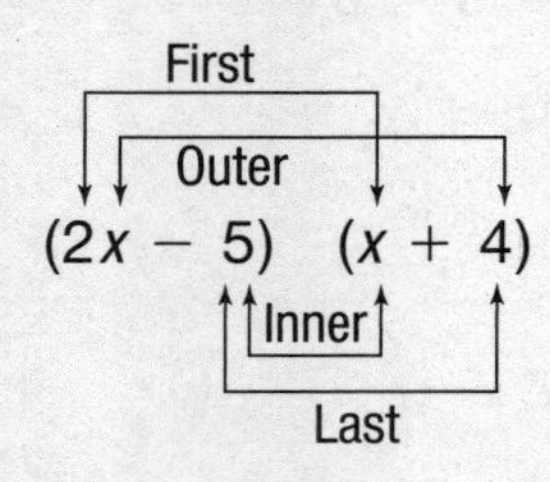

This shows how to apply the FOIL method:

Multiply: $(2x - 5)(x + 4)$	
Multiply the **F**irst terms.	$2x \cdot x = 2x^2$
Multiply the **O**uter terms.	$2x \cdot 4 = 8x$
Multiply the **I**nner terms.	$-5 \cdot x = -5x$
Multiply the **L**ast terms.	$-5 \cdot 4 = -20$
Add the products. Combine like terms.	$2x^2 + 8x - 5x - 20$
	$2x^2 + 3x - 20$

EXAMPLE 3

Use FOIL to multiply: $(3x - 2)(x + 6)$

STRATEGY **Use the FOIL method to multiply two binomials.**

STEP 1 Multiply the First terms.

$3x \cdot x = 3x^2$

STEP 2 Multiply the Outer terms.

$3x \cdot 6 = 18x$

STEP 3 Multiply the Inner terms.

$-2 \cdot x = -2x$

STEP 4 Multiply the Last terms.

$-2 \cdot 6 = -12$

STEP 5 Add the products and combine like terms.

$3x^2 + 18x - 2x - 12 = 3x^2 + 16x - 12$

SOLUTION $\mathbf{(3x - 2)(x + 6) = 3x^2 + 16x - 12}$

COACHED EXAMPLE

Simplify: $(3x + 1)^2$

THINKING IT THROUGH

The binomial has an exponent of _______, so write the binomial as a factor 2 times.

(____________) (____________)

The First terms are _______ and _______.

The product of the first terms is _______.

The Outer terms are _______ and _______.

The product of the outer terms is _______.

The Inner terms are _______ and _______.

The product of the inner terms is _______.

The Last terms are _______ and _______.

The product of the last terms is _______.

Add the four products.

$(3x + 1)^2 =$ _______________

Lesson Practice

Choose the correct answer.

1. Multiply: $m(m + 12)$

A. $2m + 12$

B. $2m + 12m$

C. $m^2 + 12$

D. $m^2 + 12m$

2. Simplify: $3x(2x - 5)$

A. $3x^2 - 15$

B. $3x^2 - 15x$

C. $6x^2 - 8x$

D. $6x^2 - 15x$

3. Simplify the expression below.

$4x(3x - 8)$

A. $12x - 32$

B. $4x^2 - 32x$

C. $12x^2 - 32$

D. $12x^2 - 32x$

4. Simplify: $(x + 9)(x + 8)$

A. $x^2 + x + 17$

B. $x^2 + 17x + 17$

C. $x^2 - 17x + 72$

D. $x^2 + 17x + 72$

5. Multiply: $(a + 6)(a - 7)$

A. $a^2 - a - 42$

B. $a^2 + a - 42$

C. $a^2 - 13a - 42$

D. $a^2 + 13a - 42$

6. Multiply: $(b - 5)(b - 8)$

A. $b^2 + 13b + 40$

B. $b^2 + 13b - 40$

C. $b^2 - 13b + 40$

D. $b^2 - 13b - 40$

7. Simplify: $(n + 6)(n - 6)$

A. $n^2 - 36$

B. $n^2 + 36$

C. $n^2 + 12n - 36$

D. $n^2 - 12n + 36$

8. Multiply.

$(x - 6)(x + 9)$

A. $x^2 + 3x - 54$

B. $x^2 - 3x + 54$

C. $x^2 - 3x - 54$

D. $x^2 + 3x + 54$

9. Simplify.

$(x + 7)(4x - 9)$

A. $4x^2 - 19x - 63$
B. $4x^2 + 19x - 63$
C. $4x^2 - 37x - 63$
D. $4x^2 + 37x - 63$

10. Multiply.

$(4d + 5)(4d + 5)$

A. $16d^2 + 25$
B. $16d^2 + 20d + 25$
C. $16d^2 + 40d + 25$
D. $16d^2 + 41d + 25$

11. Simplify the expression below.

$(4x - 3)^2$

A $16x^2 - 9$
B $16x^2 + 9$
C $16x^2 - 24x + 9$
D $16x^2 - 24x - 9$

12. Multiply.

$(x + 9)(2x + 3)$

A. $x^2 + 21x + 27$
B. $2x^2 + 16x + 27$
C. $2x^2 + 18x + 27$
D. $2x^2 + 21x + 27$

13. Multiply.

$-8n(n - 3)$

Answer ______________

14. Simplify.

$(2x - 5)(3x + 1)$

Answer ______________

Lesson

12 Factor Polynomials

Getting the Idea

When you **factor** a polynomial, you write the polynomial as the product of other polynomials. The first step in factoring a polynomial is to find the greatest common factor (GCF) of all the terms. The GCF of two or more monomials is the product of the GCF of the coefficients (and/or constants) and the GCF of the variable factors.

For example, the polynomial $12n^2 + 24n - 18$ consists of three monomials. The coefficients (12 and 24) and the constant (18) have a GCF of 6. The variable n appears in only two of the three monomials, so the polynomial has no common variable factor.

To write the polynomial in factored form using the GCF, write the common factor outside parentheses and the remaining factors inside:

$$12n^2 + 24n - 18 = 6(2n^2 + 4n - 3)$$

You can check your factoring by using the distributive property.

$$\begin{aligned} 6(2n^2 + 4n - 3) &= (6 \cdot 2n^2) + (6 \cdot 4n) - (6 \cdot 3) \\ &= 12n^2 + 24n - 18 \end{aligned}$$

EXAMPLE 1

Factor: $5x + 20y$

STRATEGY **Find the greatest common factor.**

STEP 1 Look at the coefficients of the terms.

The coefficients are 5 and 20.

The greatest common factor of 5 and 20 is 5.

STEP 2 Look at the variables.

The variables are x and y.

There is no common variable factor.

STEP 3 Factor using the greatest common factor.

Write the common factor outside parentheses and the remaining factors inside.

$5(x + 4y)$

SOLUTION $\mathbf{5x + 20y = 5(x + 4y)}$

EXAMPLE 2

Factor: $4x^4 - 6x^2$

STRATEGY **Find the GCF.**

STEP 1 Look at the coefficients of the terms.

The coefficients are 4 and 6.

The GCF of 4 and 6 is 2.

STEP 2 Look at the variables.

Both terms have x variables.

The GCF of x^4 and x^2 is x^2.

STEP 3 Factor using the GCF.

The GCF is $2 \cdot x^2$, or $2x^2$.

Write the GCF outside parentheses and the remaining factors inside.

$2x^2(2x^2 - 3)$

SOLUTION $\mathbf{4x^4 - 6x^2 = 2x^2(2x^2 - 3)}$

Note: Make sure you find the greatest common factor. In Example 2, $2x$ is a common factor of $4x^4 - 6x^2$, and $2x(2x^3 - 3x) = 4x^4 - 6x^2$. However, the terms inside parentheses, $2x^3$ and $3x$, still have x as a common factor. The *greatest* common factor is $2x^2$, not $2x$. Always check the terms inside parentheses for any other common factors.

EXAMPLE 3

Factor: $15x^5 + 5x^3 + 25x$

STRATEGY **Find the GCF.**

STEP 1 Look at the coefficients.

The GCF of 15, 5, and 25 is 5.

STEP 2 Look at the variables.

The GCF of x^5, x^3, and x is x.

STEP 3 Factor using the GCF.

The GCF is $5 \cdot x$, or $5x$.

Write the GCF outside parentheses and the remaining factors inside.

$5x(3x^4 + x^2 + 5)$

SOLUTION $\mathbf{15x^5 + 5x^3 + 25x = 5x(3x^4 + x^2 + 5)}$

COACHED EXAMPLE

What is $2a^3b^3 - 8a^2b + 12ab^2$ in factored form?

THINKING IT THROUGH

The coefficients are ________, ________, and ________.

The greatest common factor of the coefficients is ________.

The variable parts of the terms are ______, ______, and ______.

The greatest common factor of the variable parts is ________.

So, the greatest common factor of the polynomial is _______ × _______ = _______.

Write the greatest common factor outside parentheses and the remaining factors inside.

In factored form, $2a^3b^3 - 8a^2b + 12ab^2 =$ ____________________.

Lesson Practice

Choose the correct answer.

1. What is the factored form of $6x^2 - 90$?

A. $2(3x - 45)$

B. $3(2x^2 - 15)$

C. $6(x^2 - 15)$

D. $6x(x - 15)$

2. What is the factored form of $8m^2 + 12m$?

A. $2(4m^2 + 6)$

B. $4(m^2 + 3)$

C. $4m(2m + 3)$

D. $4m(2m^2 + 3)$

3. What is the factored form of $32x^2y - 20xy^2$?

A. $4xy(8x - 5y)$

B. $4xy(8y - 5x)$

C. $4x^2y^2(8x - 5y)$

D. $4x^2y^2(8x - 5)$

4. What is the factored form of $7n^2 + 42n + 21$?

A. $7(7n^2 + 6n + 3)$

B. $7(n^2 + 6n + 3)$

C. $7n(n^2 + 6n + 3)$

D. $7n^2(n + 9)$

5. Factor using the greatest common factor (GCF):

$$8x^3 + 7x^2 - 6x$$

A. $2x(4x^2 + 7x - 3)$

B. $2(4x^3 + 5x - 3)$

C. $x(8x^2 + 7x - 6)$

D. $x^2(8x^2 + 7x - 6)$

6. Factor using the greatest common factor (GCF):

$$12c^2 - 8c + 4$$

A. $4(3c^2 - 2c)$

B. $4(3c^2 - 2c + 1)$

C. $4(3c^2 - 2c + 4)$

D. $4c(3c^2 - 2c + 1)$

7. Factor using the greatest common factor (GCF):

$$6a^4b + 6a^3b^3 + 6a^2b^2$$

A. $6ab(a^3 + a^2b^2 + ab)$

B. $6ab(a^4b + a^3b^3 + a^2b^2)$

C. $6a^2b^2(a^3 + a^2b^2 + ab)$

D. $6a^2b(a^2 + ab^2 + b)$

8. Factor using the greatest common factor (GCF):

$$15x^4 - 12x^3 + 9x^2$$

A. $x(15x^4 - 12x^3 + 9x^2)$

B. $3(5x^4 - 4x^3 + 9x^2)$

C. $3x(5x^3 - 4x^2 + 3)$

D. $3x^2(5x^2 - 4x + 3)$

9. Factor using the greatest common factor (GCF):

$24n^2 + 16n - 8nb$

Answer ______________________

10. Factor using the greatest common factor (GCF):

$3t^2 - 6t$

Answer ______________________

EXTENDED-RESPONSE QUESTION

11. Maria is trying to find the GCF of the polynomial $6x^3y^4 + 12x^5y^2 - 15x^2y^3$. The GCF she found is $6x^2y^2$.

Part A Is the GCF Maria found correct?

____

Part B Explain how you know your answer is correct.

____

____

____

Lesson

13 Factor Trinomials

Getting the Idea

A trinomial factors into the product of two binomials. When you factor a trinomial, of the form $ax^2 + bx + c$, where $a = 1$, look for two factors of c (the third term) whose sum is b (the coefficient of the second term).

EXAMPLE 1

Factor: $x^2 + 6x + 9$

STRATEGY **Look for two factors of 9 whose sum is 6.**

STEP 1 List the factor pairs of 9 and their sums until you find the correct pair.

Since the only operation in the trinomial is addition, you are looking for numbers to complete the binomials: $(x + \_\_\_)(x + \_\_\_)$

Factors of 9	Sum of Factors
1, 9	$1 + 9 = 10$ (NO)
3, 3	$3 + 3 = 6$ (YES)

STEP 2 Write $x^2 + 6x + 9$ as the product of two binomials.

$(x + 3)(x + 3)$

STEP 3 Use the FOIL method to check $(x + 3)(x + 3)$.

First: $x \cdot x = x^2$

Outer: $x \cdot 3 = 3x$

Inner: $3 \cdot x = 3x$

Last: $2 \cdot 3 = 6$

So, $(x + 3)(x + 3) = x^2 + 6x + 9$, and the factored form checks.

SOLUTION $\mathbf{x^2 + 6x + 9 = (x + 3)(x + 3)}$

Note: Here is a shortcut that can help you factor trinomials. If the only operation in a trinomial is addition (as in $x^2 + 11x + 24$), then you know that its binomial factors will be in the form $(x + m)(x + n)$, where m and n are a pair of factors of the third term in the trinomial.

When you factor a trinomial with one or more subtraction signs, you'll need to consider negative integers as possible factors.

EXAMPLE 2

Factor: $x^2 + 24x - 25$

STRATEGY **Look for two factors of −25 whose sum is 24.**

STEP 1 List the factor pairs of −25 and their sums until you find the correct pair.

Factors of −25	Sum of Factors
1, −25	1 + (−25) = −24 (NO)
−1, 25	−1 + 25 = 24 (YES)
5, −5	5 + (−5) = 0 (NO)

The factors −1 and 25 have a sum of 24, so use −1 and 25 as the second terms in the two binomials: $(x - 1)(x + 25)$

STEP 2 Use the FOIL method to check $(x - 1)(x + 25)$.

First: $x \cdot x = x^2$

Outer: $x \cdot 25 = 25x$

Inner: $-1 \cdot x = -x$

Last: $-1 \cdot 25 = -25$

Add the products. Combine like terms.

$(x - 1)(x + 25) = x^2 + 25x - x - 25 = x^2 + 24x - 25$

The factored form checks.

SOLUTION $\mathbf{x^2 + 24x - 25 = (x - 1)(x + 25)}$

COACHED EXAMPLE

Factor: $x^2 - 8x + 7$

THINKING IT THROUGH

Look for two numbers whose product is _____ and whose sum is −_______.

Make a list of the factors of the third term. Check if the sum of those factors is the coefficient of the second term.

Factors of 7	Sum of Factors
1, 7	_______ + _______ = _______
	_______ + _______ = _______

The factors _____ and _______ have a sum of ______.

The binomial factors of $x^2 - 8x + 7$ are (_________) and (_________).

The factored form of $x^2 - 8x + 7$ is ________________.

Lesson Practice

Choose the correct answer.

1. Factor: $x^2 + 4x + 3$

A. $(x + 1)(x + 3)$
B. $(x + 1)(x - 3)$
C. $(x - 1)(x - 3)$
D. $(x - 1)(x + 3)$

2. Factor: $x^2 + 12x + 11$

A. $(x + 1)(x + 11)$
B. $(x + 2)(x + 6)$
C. $(x + 3)(x + 4)$
D. $(x + 12)(x - 1)$

3. Factor: $x^2 - 5x + 6$

A. $(x + 1)(x + 6)$
B. $(x - 1)(x - 6)$
C. $(x + 2)(x + 3)$
D. $(x - 2)(x - 3)$

4. Factor: $x^2 + 14x + 13$

A. $(x - 1)(x - 14)$
B. $(x + 1)(x + 13)$
C. $(x + 14)(x - 1)$
D. $(x + 2)(x + 7)$

5. Factor: $x^2 + 4x - 5$

A. $(x + 5)(x - 1)$
B. $(x - 5)(x + 1)$
C. $(x + 5)(x + 1)$
D. $(x - 5)(x - 1)$

6. Factor: $x^2 + 9x + 18$

A. $(x - 3)(x - 6)$
B. $(x + 9)(x + 2)$
C. $(x + 3)(x + 6)$
D. $(x + 1)(x + 18)$

7. Factor: $x^2 - x - 6$

A. $(x - 3)(x - 2)$
B. $(x + 2)(x - 3)$
C. $(x + 3)(x - 2)$
D. $(x - 1)(x + 6)$

8. Factor: $x^2 + 10x + 16$

A. $(x - 4)(x - 4)$
B. $(x + 4)(x + 4)$
C. $(x - 2)(x - 8)$
D. $(x + 2)(x + 8)$

9. What is $x^2 + 5x + 6$ in factored form?

Answer ____________________

10. What is $x^2 + 7x + 10$ in factored form?

Answer ____________________

Lesson

14 Represent Numerical Information

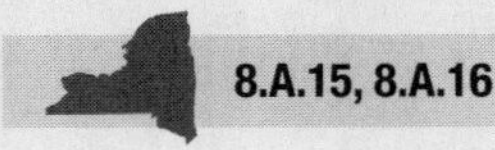

Getting the Idea

A data set can be represented in a variety of ways. For example, consider the data set represented in the table below.

x	−2	−1	0	1	2
y	−1	0	1	2	3

This data set can also be represented *verbally*: *y* is 1 more than *x*.

This data set can also be represented *algebraically*: $y = x + 1$.

This data set can also be represented *graphically*:

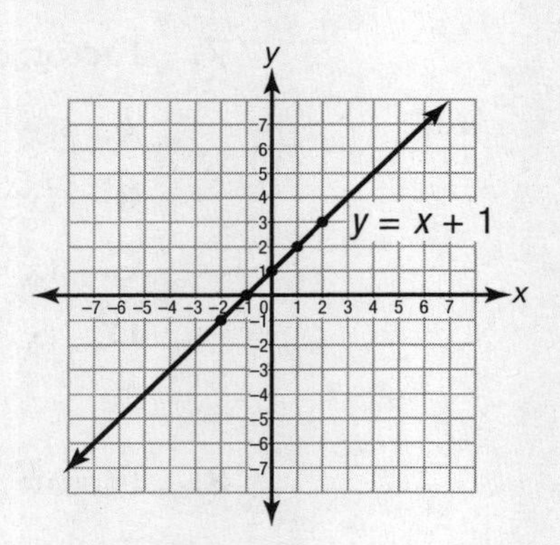

EXAMPLE 1

A video game rents for $6.00 per day. Tyler needs to rent a cable for his video game player in order to play the game. The cable rental is $1.50 per day. What is an equation that represents the amount in dollars, *y*, that it costs to rent the game for *x* days?

STRATEGY **Find the rule that relates *x* to *y*.**

STEP 1 Find the total cost per day.

The game costs $6.00 per day, and the cable costs $1.50 per day.

Total cost per day: $6.00 + 1.50 = 7.50$, or $7.50 per day

STEP 2 Find the total cost for *x* days.

If the cost for one day is $7.50, then the cost for *x* days is $7.50x$, or $7.5x$.

STEP 3 Find the relationship between x and y.

$y = 7.5x$

SOLUTION **The equation that represents the amount in dollars, y, that it costs Tyler to rent the game for x days is $y = 7.5x$.**

EXAMPLE 2

The graph below shows the number of miles traveled by a car and the amount of gas it has used. Represent the data with an equation.

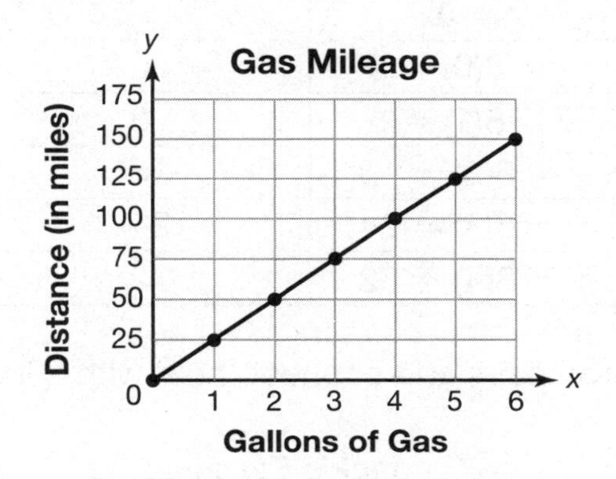

STRATEGY **Make a table to find the rule. Then use the rule to write an equation.**

STEP 1 Make a table.

Gas Mileage

Gallons of gas (x)	0	1	2	3	4	5	6
Distance in miles (y)	0	25	50	75	100	125	150

STEP 2 Find the rule from the table.

The number of miles increases by 25 with each gallon, so the number of miles is 25 times the number of gallons.

STEP 3 Represent the rule with an equation.

$y = 25x$

SOLUTION **The equation $y = 25x$ represents the data in the graph.**

EXAMPLE 3

Jack works as a cashier in a grocery store. He earns $8 per hour. An equation that represents this situation is $y = 8x$, where x is the number of hours he works and y is the amount of money he earns. Represent this situation in a graph.

STRATEGY **Make a table of values to find ordered pairs using the equation. Then graph the ordered pairs.**

STEP 1 Make a table of values.

x	$y = 8x$	y	(x, y)
0	$y = 8(0) = 0$	0	(0, 0)
1	$y = 8(1) = 8$	8	(1, 8)
2	$y = 8(2) = 16$	16	(2, 16)
3	$y = 8(3) = 24$	24	(3, 24)
4	$y = 8(4) = 32$	32	(4, 32)

STEP 2 Graph the ordered pairs and connect them with a line.

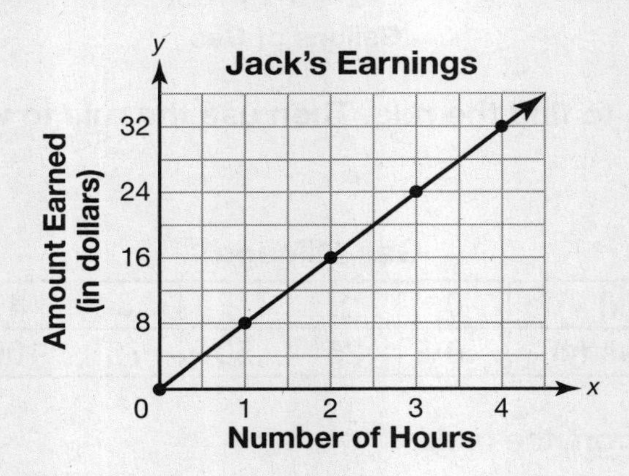

SOLUTION **The graph that represents the situation is shown in Step 2.**

COACHED EXAMPLE

Represent the equation $y = -\frac{1}{3}x$ with a graph.

THINKING IT THROUGH

Make a table of values. Use the x-values shown in the first column of the table.

x	$y = -\frac{1}{3}x$	(x, y)
−6	$y = -\frac{1}{3} \times (-6) =$ ____	( )
−3		( )
0		( )
3		( )
6		( )

Plot and label the ordered pairs on the coordinate plane below. Then connect the points.

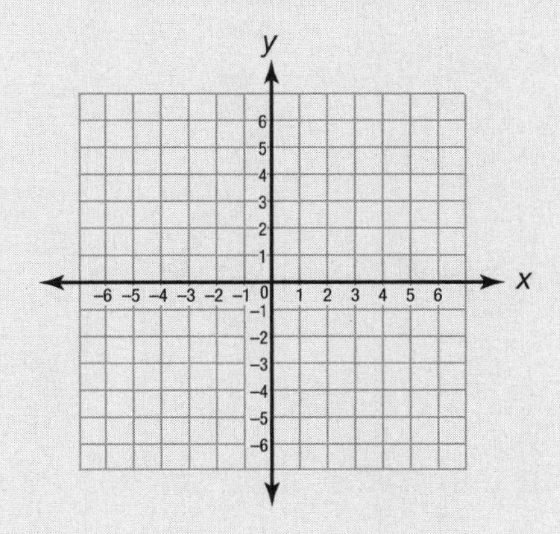

The graph above represents the equation $y = -\frac{1}{3}x$.

Lesson Practice

Choose the correct answer.

1. Each student in a ceramics class needs to buy a bag of clay that costs \$4.90 and a small trowel that costs \$3.00. Let x represent the number of students in the class and y represent the total amount of money spent by the class. Which equation represents this situation?

A. $y = 4.9x + 3$

B. $y = 7.9x$

C. $y = x + 4.9$

D. $y = x + 7.9$

2. This table shows a relationship between x and y.

x	y
0	5
2	9
3	11
6	17
7	19

Which equation best represents these data?

A. $y = x + 5$

B. $y = 5x$

C. $y = 4x + 1$

D. $y = 2x + 5$

3. Which graph represents the equation $y = -\frac{1}{2}x$?

A.

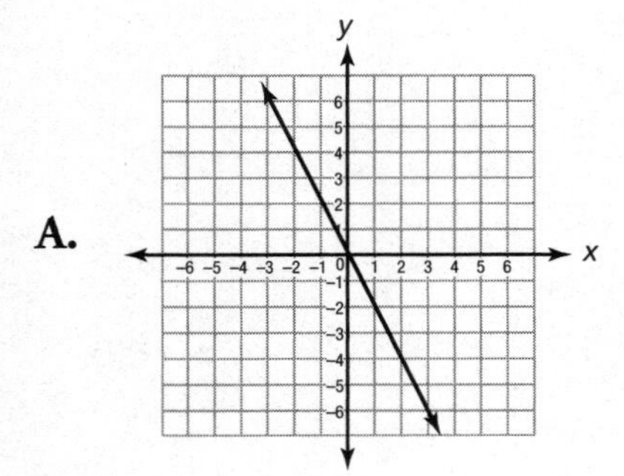

B.

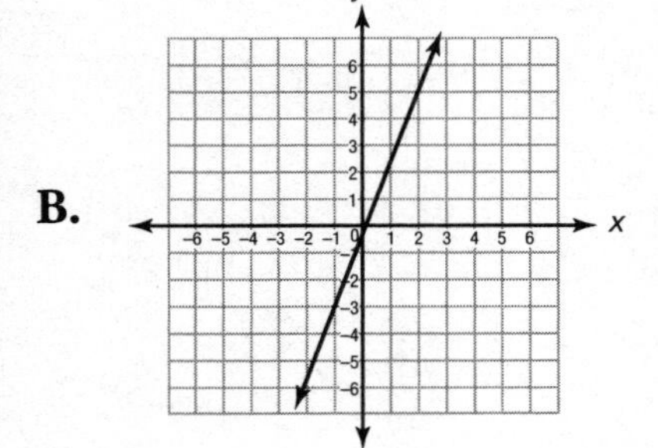

C.

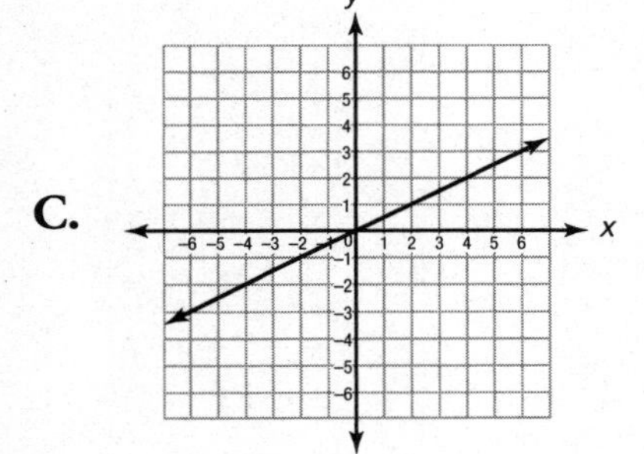

D.

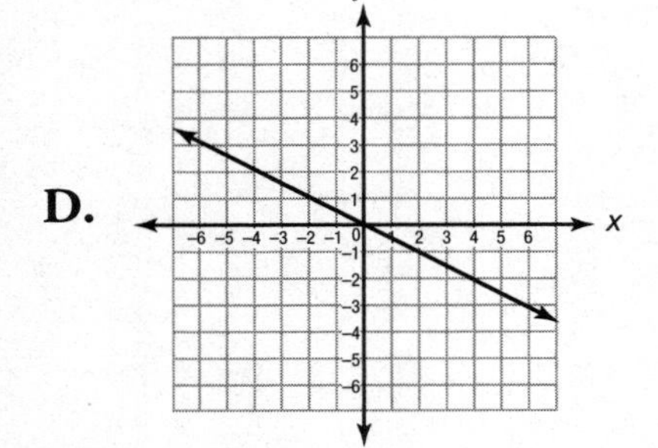

4. Which of the following is the best verbal representation of the data in this table?

x	−2	0	2	4	6
y	10	0	−10	−20	−30

A. Multiply each x-value by −5 to get each y-value.

B. Multiply each x-value by 5 to get each y-value.

C. Subtract 8 from each x-value to get each y-value.

D. Add 8 to each x-value to get each y-value.

5. Which equation is an algebraic representation of the data shown in this graph?

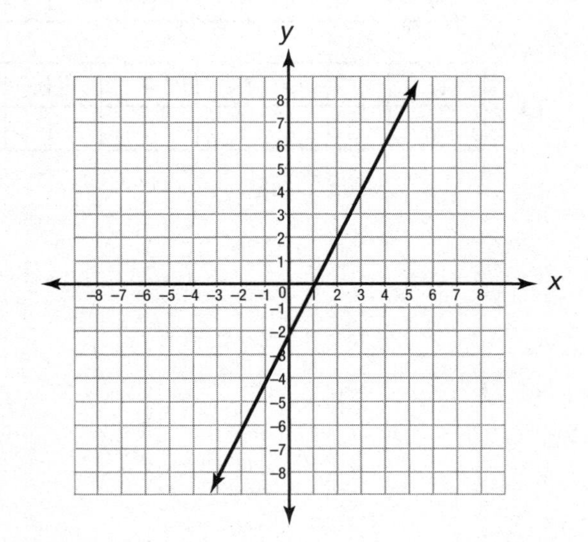

A. $y = 2x + 2$

B. $y = 2x - 2$

C. $y = -2x + 2$

D. $y = -2x - 2$

6. Which table of values represents the relationship shown in the graph?

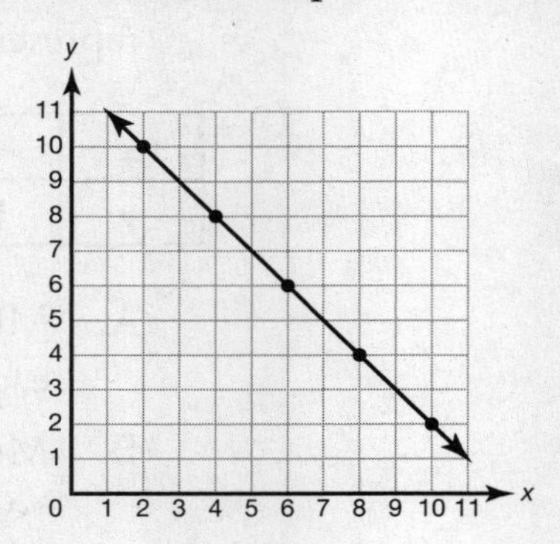

A.

x	10	8	6	4	2
y	2	4	6	8	10

B.

x	2	4	6	8	10
y	2	4	6	8	10

C.

x	2	4	6	8	10
y	10	8	6	4	2

D.

x	2	6	4	10	8
y	10	8	6	4	2

EXTENDED-RESPONSE QUESTION

7. Use the equation $y = -2x + 1$ for Parts A and B.

Part A

Complete this table. Then list the ordered pairs from the table.

x	-3	-1	0	1	3
y					

Ordered pairs: ____

Part B

Graph the ordered pairs you found in Part A and draw the line.

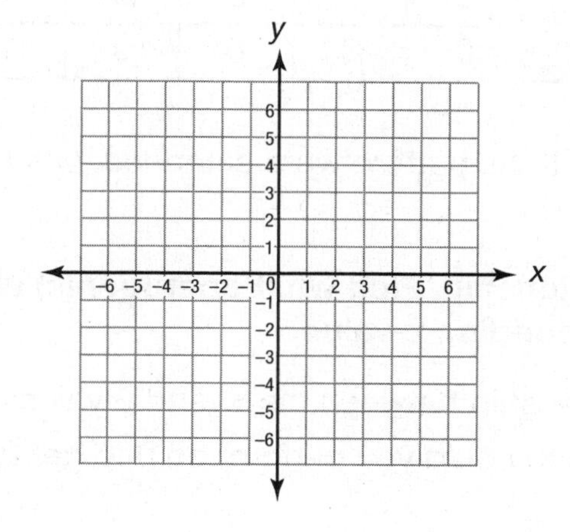

Lesson

15 Use Patterns to Develop Rules

Getting the Idea

A pattern is an arrangement of numbers or objects. Pairs of numbers can form patterns.

EXAMPLE 1

Look at the numbers in the table below.

x	0	1	2	3	4
y	0	3	6	9	?

What is a rule for this pattern? If the pattern were extended, what is the y-value when the x-value is 4?

STRATEGY **Look for a relationship. Find what operation(s) were used to get a y-value from its corresponding x-value.**

STEP 1 Look for a relationship between the x- and y-values.

What operation could you perform on 0 to get 0? Add 0, or multiply by any number.

What operation could you perform on 1 to get 3? Add 2, or multiply by 3.

What operation could you perform on 2 to get 6? Add 4, or multiply by 3.

What operation could you perform on 3 to get 9? Add 6, or multiply by 3.

What operation could you perform on 4 to get 12? Add 8, or multiply by 3.

STEP 2 Find the rule that is the same for all pairs of numbers.

For each pair of values, you could multiply the x-value by 3 to get the y-value. The rule for the pattern is to multiply the x-value by 3, which can be represented by the equation $y = 3x$.

STEP 3 Find the y-value when the x-value is 4.

$y = 3x$

$y = 3 \times 4 = 12$

SOLUTION **The rule is to multiply by 3, or $y = 3x$. When $x = 4$, $y = 12$.**

You can use patterns to find a formula for determining the sum of the measures of the interior angles of a polygon.

A diagonal is a line segment that joins two non-adjacent vertices of a polygon. You can draw diagonals in polygons with four or more sides to divide the figure into triangles. Remember that the sum of the angles of a triangle is 180°.

EXAMPLE 2

What is the formula (or rule) for finding the sum of the measures of the interior angles of a polygon?

STRATEGY **Divide polygons into triangles by drawing all of the diagonals from one vertex.**

STEP 1 Make a chart showing the names of polygons, the number of sides, and sketches of the figures.

Polygon	Number of Sides	Sketch
Quadrilateral	4	
Pentagon	5	
Hexagon	6	
Heptagon	7	
Octagon	8	

STEP 2 Draw all the diagonals from one vertex in each polygon. Count the number of triangles formed.

Polygon	Number of Sides	Polygon Divided into Triangles	Number of Triangles
Quadrilateral	4		2
Pentagon	5		3
Hexagon	6		4
Heptagon	7		5
Octagon	8		6

STEP 3 Describe the relationship between the number of sides and the number of triangles formed.

For each polygon, the number of triangles formed is 2 less than the number of sides.

STEP 4 Find the sum of the measures of the interior angles of each polygon.

There are 180° in each triangle, so multiply 180° by the number of triangles in each polygon.

Polygon	Number of Sides (n)	Number of Triangles ($n - 2$)	Sum of the Interior Angles $(n - 2)180$
Quadrilateral	4	2	$2 \times 180 = 360$
Pentagon	5	3	$3 \times 180 = 540$
Hexagon	6	4	$4 \times 180 = 720$
Heptagon	7	5	$5 \times 180 = 900$
Octagon	8	6	$6 \times 180 = 1{,}080$

The formula for the sum of the measures of the interior angles of a polygon is $(n - 2)180°$, where n is the number of sides in the polygon.

SOLUTION **The sum of the measures of the interior angles of a polygon equals $(n - 2)180°$, where n is the number of sides in the polygon.**

COACHED EXAMPLE

Ginger has her laundry done at the local laundromat. The laundromat charges a flat fee of $2.00 plus a certain cost per load. The table below shows how much it costs Ginger to do her laundry at the laundromat.

Number of loads (n)	1	2	3	4
Total cost (c)	$4.50	$7.00	$9.50	$12.00

Write an equation that expresses Ginger's total cost (c) of doing her laundry in terms of the number of loads of laundry (n) that Ginger does.

THINKING IT THROUGH

Find the change in c when n increases by 1.

c increases by ________, so c might equal ________$n + 2$ (the flat fee).

Multiply each value of n by ________, and then add _____ to see if you get the corresponding c-value.

1 × ________ + ________ = ________

2 × ________ + ________ = ________

3 × ________ + ________ = ________

4 × ________ + ________ = ________

So, c is ________ times n plus ________.

An equation that represents the situation is c = ____________________.

Lesson Practice

Choose the correct answer.

1. What is the equation for this table?

x	1	2	3	4	5
y	6	7	8	9	10

A. $y = 3x + 3$
B. $y = x + 5$
C. $y = 2x$
D. $y = 6x$

2. Which is the rule for this table?

x	2	3	4	5	6
y	7	10	13	16	19

A. $y = 4x - 1$
B. $y = 5x - 3$
C. $y = 2x - 3$
D. $y = 3x + 1$

3. A decagon is a polygon with 10 sides. Which expression can be used to find the sum of the measures of the interior angles of a decagon?

A. $(8 - 2)180$
B. $(10 - 2)180$
C. $(180 - 10)2$
D. $(12 - 2)180$

4. Dan made this table to show the distance he had traveled at the end of each hour for 4 hours.

Time, t	1	2	3	4	5
Distance, d	55	110	165	220	275

Which shows an equation for the distance Dan traveled?

A. $d = 55t$
B. $55 = dt$
C. $t = 55d$
D. $d = \frac{t}{55}$

Use the table below for questions 5 and 6.

The table indicates how much it costs to rent a canoe at a lake. The rental company charges a flat fee of $8.00, plus a certain amount for each hour.

Hours (h)	1	2	3	4
Total cost (c)	$14	$20	$26	$32

5. Write an equation that expresses the total cost of renting a canoe (c) in terms of the number of hours (h) that it is rented.

Answer ______________________

6. How much will it cost to rent the canoe for 6 hours?

Answer ______________________

Lesson

16 Relations

Getting the Idea

A **relation** is any set of ordered pairs and is represented as (x, y). A relation may or may not have ordered pairs that have the same x-value.

You can represent relations using tables, graphs, and equations.

EXAMPLE 1

Write a rule for the relation shown in the table.

x	2	4	6	8	10
y	8	16	24	32	40

STRATEGY **Look for a pattern in the way each x-value relates to its corresponding y-value.**

STEP 1 Look at each x-value and its corresponding y-value. Describe the relationship in words.

Each y-value is 4 times its corresponding x-value.

STEP 2 Write an equation using x and y.

$y = 4x$

SOLUTION **An equation to represent the relation shown in the table is $y = 4x$.**

EXAMPLE 2

Make a table of values to represent the relation shown in the graph.

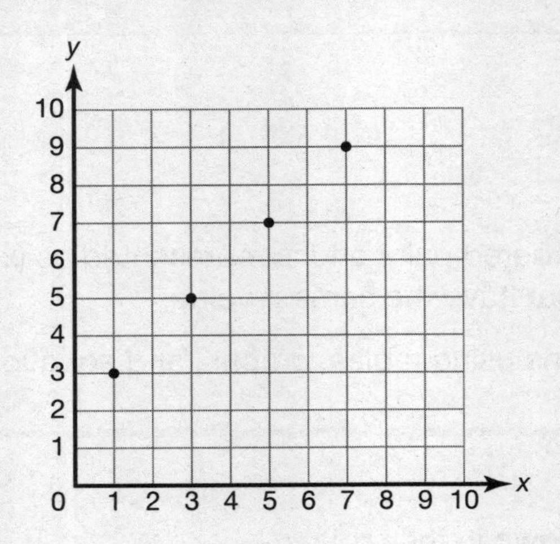

STRATEGY **Write the ordered pairs.**

STEP 1 Write the ordered pairs.

(1, 3) (3, 5) (5, 7) (7, 9)

STEP 2 Enter the x-values and the corresponding y-values into the table.

x	1	3	5	7
y	3	5	7	9

SOLUTION **A table that represents the relation is shown in Step 2.**

COACHED EXAMPLE

Write the equation for the relation shown in the table.

x	5	6	7	8	9	10
y	10	11	12	13	14	15

THINKING IT THROUGH

Each y-value is ________ more than its corresponding x-value.

The equation $y =$ ________ + ________ represents the relation.

An equation that represents the relation shown in the table is $y =$ ____________.

Lesson Practice

Choose the correct answer.

1. Jill earns bonus points every time she makes it to the next level on a video game. The table below shows how many points, *y*, she earns for each of the levels, *x*, she makes it to. Which graph represents the relation in the table?

x	1	2	3	4
y	2	4	7	9

A. 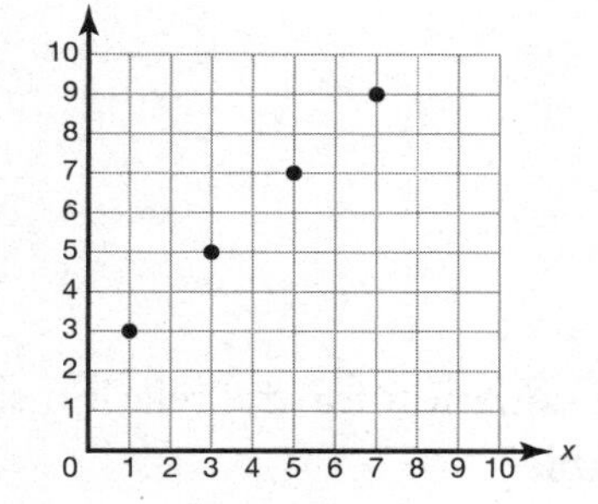

C. 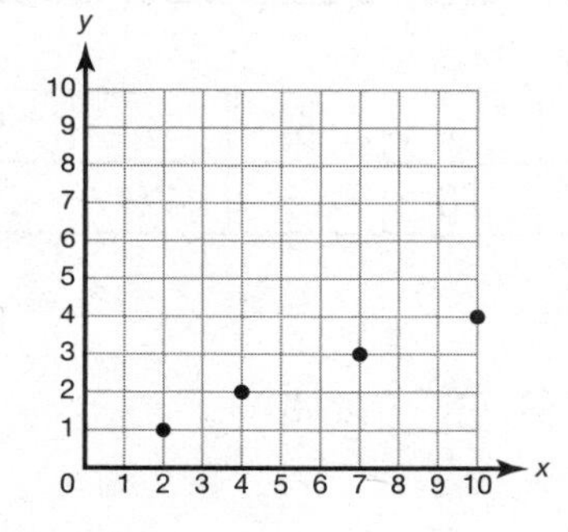

B. 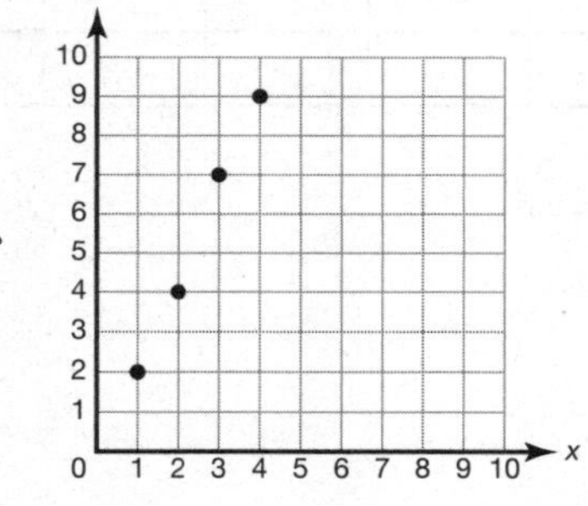

D. 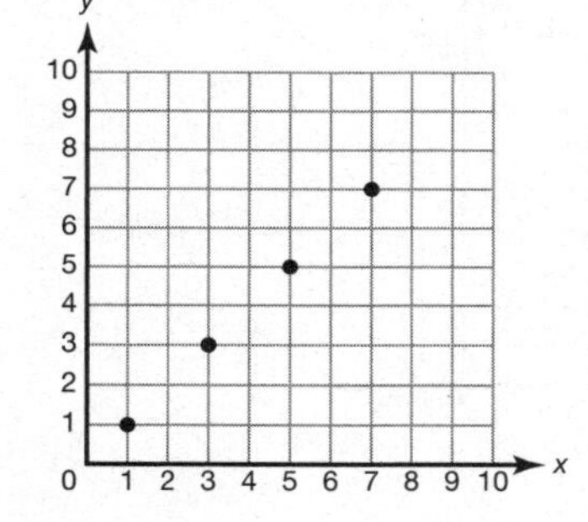

2. Which is the equation for the relation shown in this table?

x	4	6	8	10	12
y	20	30	40	50	60

A. $y = 5x$

B. $y = x \div 5$

C. $y = x - 5$

D. $y = x + 5$

3. Which equation represents the relation shown in the graph?

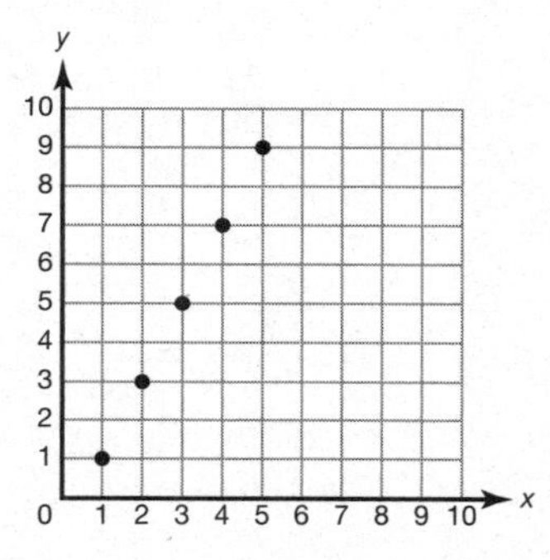

A. $y = x + 1$

B. $y = 2x - 1$

C. $y = 3x$

D. $y = 3x - 2$

4. Which table represents $y = 3x + 2$?

A.

x	1	3	5	7	9
y	5	11	17	23	29

B.

x	1	3	5	7	9
y	3	6	9	12	15

C.

x	1	3	5	7	9
y	5	7	9	11	25

D.

x	1	3	5	7	9
y	5	10	15	20	25

5. Write an equation for this graph.

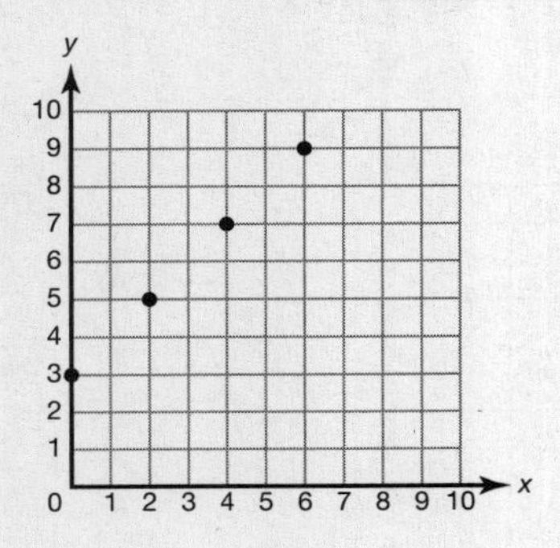

Answer ____________

6. Write an equation for the relation shown in this table.

x	2	4	6	8	10
y	6	10	14	18	22

Answer ____________

Lesson

17 Identify Functions

8.A.17, 8.A.18

Getting the Idea

A **function** is a relation in which each x-value has only one y-value.

The **domain** of a function is the set of all the possible input values, or x-values, in a given situation. The **range** of a function is the set of all the possible output values, or y-values, in a given situation.

EXAMPLE 1

Give the domain and range for the function shown below.

Sale Prices

Regular Price, x	$5	$10	$15	$20	$25
Sale Price, y	$1	$2	$3	$4	$5

STRATEGY **Identify the domain and range of the function.**

STEP 1 Identify the domain.

List the x-values.

{5, 10, 15, 20, 25}

STEP 2 Identify the range.

List the y-values.

{1, 2, 3, 4, 5}

SOLUTION **The domain of the function is {5, 10, 15, 20, 25}, and the range is {1, 2, 3, 4, 5}.**

If a relation is given as a set of ordered pairs, you can determine whether the relation is a function by looking at the *x*- and *y*-coordinates. If any ordered pairs have the same *x*-coordinate but different *y*-coordinates, the relation is not a function.

EXAMPLE 2

Is the relation represented in the table a function?

Input (*x*)	Output (*y*)
5	−2
2	−1
1	0
2	1
5	2

STRATEGY **Compare the input numbers to the output numbers. If each input number has only one output number, then the relation is a function.**

STEP 1 Compare input and output numbers.

5 has an output of −2 and 2.

2 has an output of −1 and 1.

1 has an output of 0.

STEP 2 Does any input number have more than one output number?

Both 5 and 2 have more than one output number.

SOLUTION **The relation is not a function.**

You can also determine whether a relation is a function by graphing the relation, and then doing a **vertical line test**.

EXAMPLE 3

Identify the domain and range of the relation graphed below. Then use a vertical line test to determine whether the relation is a function.

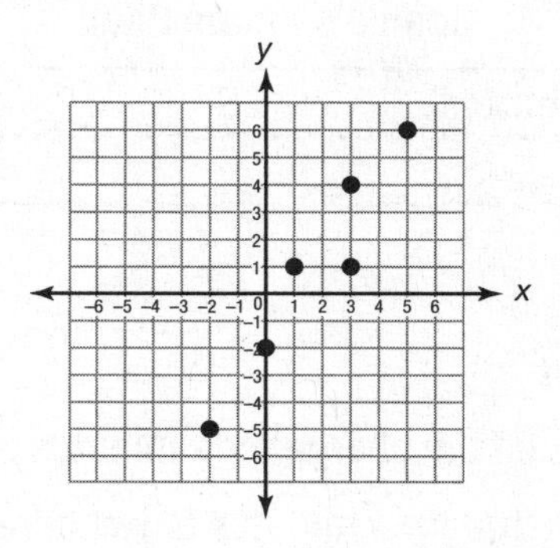

STRATEGY **Identify the domain and range. Then do a vertical line test.**

STEP 1 Identify the ordered pairs.

The ordered pairs are $(-2, -5)$, $(0, -2)$, $(1, 1)$, $(3, 1)$, $(3, 4)$, and $(5, 6)$.

STEP 2 Identify the domain and range.

The domain consists of the set of x-values: $\{-2, 0, 1, 3, 5\}$

The range consists of the set of y-values: $\{-5, -2, 1, 4, 6\}$

STEP 3 Do a vertical line test to determine whether the relation is a function.

First create the graph of the relation by drawing a line to connect the points.

Then determine whether any vertical line intersects that graph at more than one point.

A vertical line intersects the graph of the relation at both $(3, 1)$ and $(3, 4)$, so the relation is not a function.

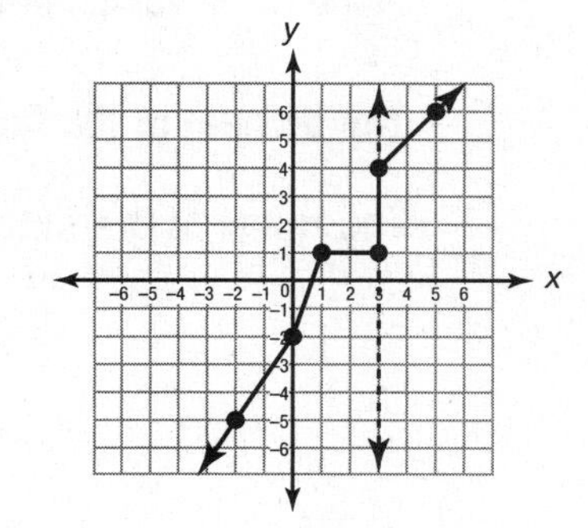

SOLUTION **The domain of the relation is $\{-2, 0, 1, 3, 5\}$.**
The range of the relation is $\{-5, -2, 1, 4, 6\}$. The relation is not a function since a vertical line intersects the graph of the relation at more than one point.

COACHED EXAMPLE

Joanie's international long-distance phone plan charges her a $2 connection fee, plus $3 for each minute for each call. Complete the function table below to determine the cost for a call of each length. Then identify the domain and range of the function.

Joanie's Phone Plan

Length of call, *x* (in minutes)	5	10	15	20	25
Cost, *y* (in dollars)					

THINKING IT THROUGH

Find an equation for the function.

The charge for one minute is $3, so the charge for x minutes is _____ dollars.

Add the connection fee to the charge for x minutes to write the equation.

$y =$ _________ + _________

Substitute values of x from the table in the rule.

If $x = 5$, $y =$ ________________________.

If $x = 10$, $y =$ ________________________.

If $x = 15$, $y =$ ________________________.

If $x = 20$, $y =$ ________________________.

If $x = 25$, $y =$ ________________________.

Place the y-values in the table.

The domain of the function is: {_____________}.

The range of the function is: {_____________}.

Lesson Practice

Choose the correct answer.

1. Which relation is a function?

A.

x	−1	0	1	2	2
y	3	4	5	6	7

B.

x	−5	−5	10	15	20
y	9	8	7	6	5

C.

x	4	8	12	16	20
y	−7	7	−7	7	−7

D.

x	6	6	6	6	6
y	−3	−2	−1	2	1

2. Which relation is not a function?

A. $\{(2, 3), (4, 5), (6, 7), (8, 9)\}$

B. $\{(1, -5), (2, -5), (3, -5), (4, -5)\}$

C. $\{(-4, -4), (-3, -2), (-2, -1), (-1, 0)\}$

D. $\{(6, 4), (7, 5), (8, 6), (8, 7)\}$

3. What is the domain of the function graphed below?

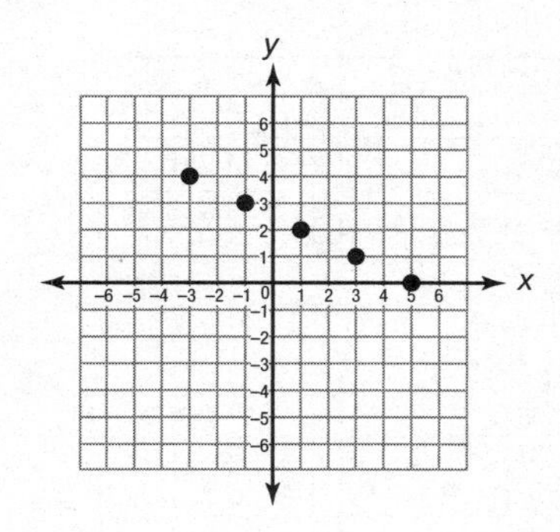

A. $\{0, 1, 2, 3, 4\}$

B. $\{-1, -3, -2, -1, 0\}$

C. $\{-3, -1, 1, 3, 5\}$

D. $\{-3, 3, 2, 1, 0\}$

4. What is the range of this relation?

x	−3	−2	−1	0	1	2	3
y	−10	−7	−4	−1	2	5	8

A. 6

B. 18

C. $\{-3, -2, -1, 0, 1, 2, 3\}$

D. $\{-10, -7, -4, -1, 2, 5, 8\}$

5. Which relation is a function?

A.

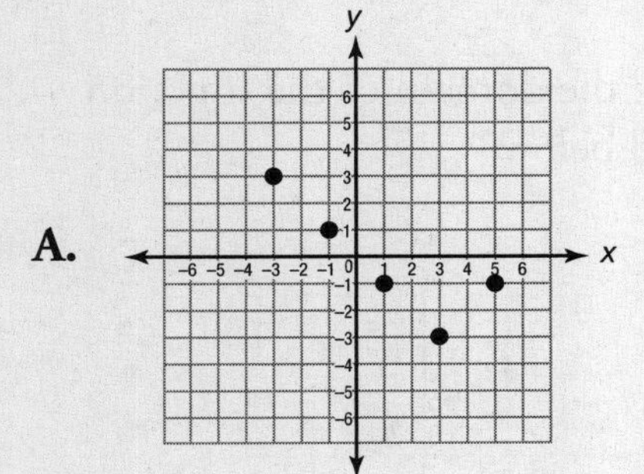

B.

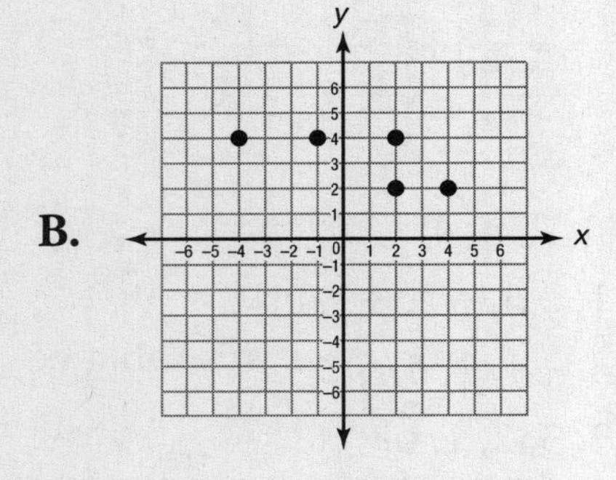

C.

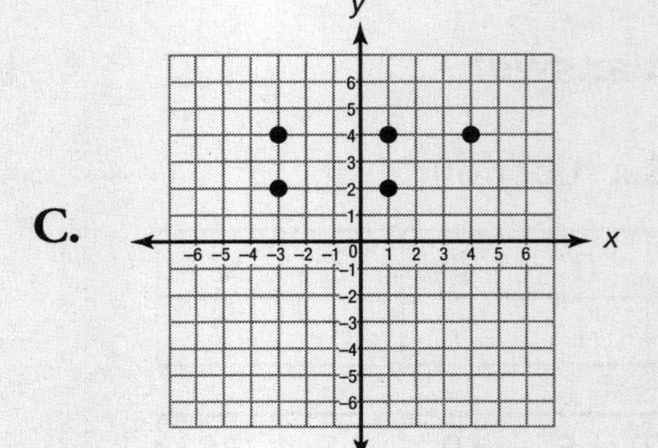

D.

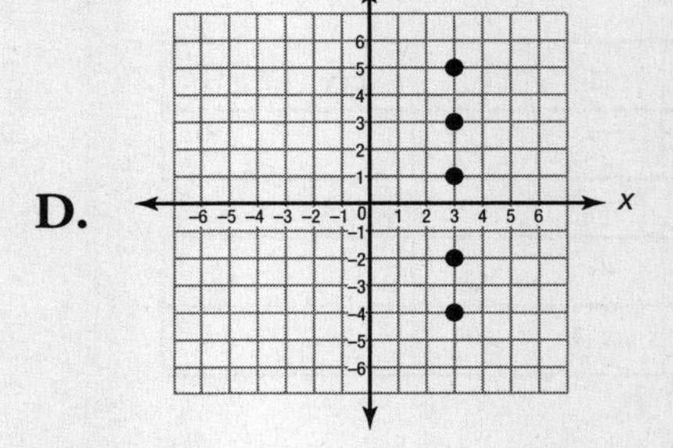

6. A function has the rule $y = 2x - 3$ and the domain $\{-2, -1, 0, 1, 2\}$. What is the range?

Answer ____________________

2 Review

1 A salesman at an appliance store earns a base salary of $550 per week plus a 6% commission on his sales. If x represents the amount he made in sales last week, which expression represents his total earnings for last week?

A $550x + 0.06$

B $550x - 0.06$

C $0.06x + 550$

D $0.06x - 550$

2 Gene is buying a cake and paper plates for a class party. The cake costs $24 and each package of paper plates costs $3. He cannot spend more than $35. Which inequality could be used to determine p, the possible number of packages of plates he could buy?

A $3p + 24 \leq 35$

B $3p - 24 \leq 35$

C $24 - 3p \leq 35$

D $3p + 24 \geq 35$

3 What is the solution of this inequality?

$$6 - 4x > 4 - 3x$$

A $x < -2$

B $x > -2$

C $x < 2$

D $x > 2$

4 Which equation does this table of values represent?

x	-1	0	1	2	3
y	-8	-3	2	7	12

A $y = x - 7$

B $y = 3x - 5$

C $y = 4x - 4$

D $y = 5x - 3$

5 The domain of the function below is $\{-2, -1, 0, 1, 2\}$.

$$y = 4x + 1$$

What is the range of the function?

A $\{-6, -4, -2, 0, 2\}$

B $\{-7, -3, 1, 5, 9\}$

C $\{-8, -4, 0, 4, 8\}$

D $\{-9, -5, -1, 3, 7\}$

6 Which relation is a function?

A $\{(-1, 1), (2, 1), (2, -2), (4, 4)\}$

B $\{(0, 5), (-1, 3), (-1, 1), (-2, -1)\}$

C $\{(11, 33), (8, 24), (-6, 30), (-9, 18)\}$

D $\{(4, 17), (-1, 11), (2, 8), (4, 5)\}$

7 Which equation represents the table below?

x	y
2	5
4	9
7	15
9	19

A $y = x + 3$

B $y = 4x - 3$

C $y = 2x + 1$

D $y = 3x - 1$

8 Which situation could be represented by this graph?

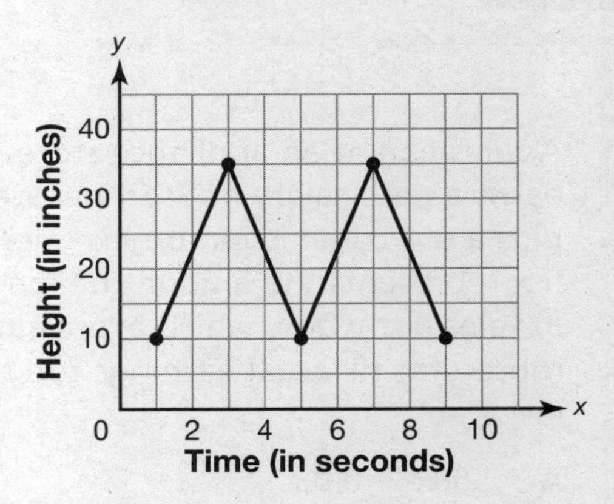

A a ride on a seesaw

B a ride on a roller coaster

C a fall down a hill

D a run on a treadmill

9 Simplify the expression below.

$$(12n - 16) - (5n + 6)$$

A $7n - 10$

B $7n - 22$

C $17n - 10$

D $17n - 22$

10 What is $48x^3$ divided by $6x$?

A $8x^2$

B $8x^3$

C $42x^2$

D $42x^3$

11 Factor the expression below.

$x^2 + 7x + 12$

A $(x + 2)(x + 6)$

B $(x + 3)(x + 4)$

C $(x - 3)(x - 4)$

D $(x - 2)(x - 6)$

12 Which equation represents the data in this table?

x	−6	−3	0	3	6
y	−3	0	3	6	9

A $y = 3x$

B $y = x + 3$

C $y = x - 3$

D $y = x \div 2$

13 Which set of ordered pairs satisfies this equation?

$y = 3x - 4$

A $\{(-2, -10), (-1, -7), (0, -4), (1, -1), (2, 2)\}$

B $\{(-2, 2), (-1, 1), (0, -4), (1, 1), (2, -2)\}$

C $\{(-2, 10), (-1, 7), (0, -4), (1, -1), (2, 2)\}$

D $\{(-2, -10), (-1, -7), (0, -4), (1, 1), (2, -2)\}$

14 What is the sum of the polynomials modeled below?

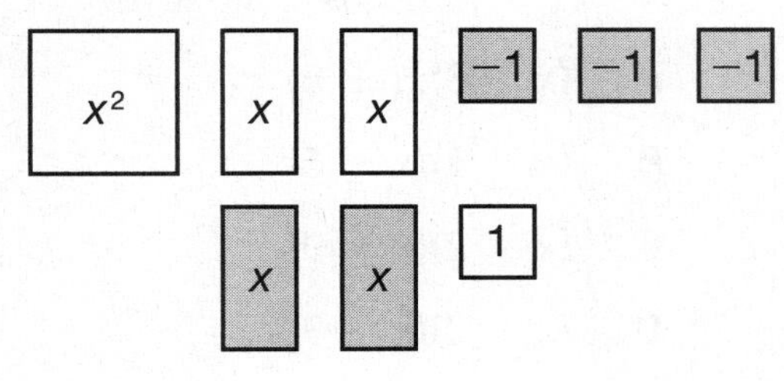

A $x^2 + 4x + 4$

B $x^2 + 2x - 2$

C $x^2 + 2$

D $x^2 - 2$

15 Multiply.

$(-3xy^2)(4x^2y)$

A $12x^2y^2$

B $-12x^2y^2$

C $-12x^3y^3$

D $-12x^6y^6$

16 Factor the expression below using the greatest common factor (GCF).

$18x^2 + 27x$

A $3(6x^2 + 9)$

B $3x(6x^2 + 3x)$

C $9x(2x + 3)$

D $9x^2(2x + 3)$

17 Factor the expression below.

$x^2 + 9x + 18$

A $(x + 3)(x + 6)$

B $(x - 3)(x - 6)$

C $(x + 2)(x + 9)$

D $(x - 2)(x - 9)$

18 Which sentence is an algebraic representation of this verbal sentence?

y is 5 less than 3 times *x*.

A $y < 3x - 5$

B $y = 5 - 3x$

C $y = 3x - 5$

D $y = 5x - 3$

19 Simplify the expression below.

$$\frac{30n^3 - 20n^2 + 25n}{-5n}$$

Show your work.

Answer ______________

20 A rectangle has a length of $(2x - 1)$ units and a width of $(x + 3)$ units. What is the perimeter of the rectangle?

Show your work.

Answer __________

21 Multiply.

$(2x + 5)(2x - 5)$

Show your work.

Answer ______________

22 Marcia earns $12 per hour at her job.

Part A

Create a graph of Marcia's earnings.

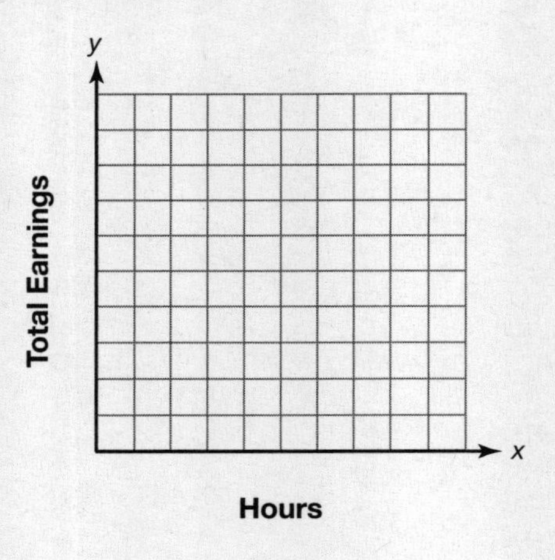

Part B

Write an equation to represent the relationship between Marcia's total earnings, y, and the number of hours she works, x.

Answer _______________

23 Use this equation.

$y = 2x + 3$

Part A

Describe the equation in words.

____

Complete this table. Then list the ordered pairs from the table.

x	−4	−2	0	2	4
y					

Answer _______________

Part B

Graph the ordered pairs you found in *Part A* and draw the line.

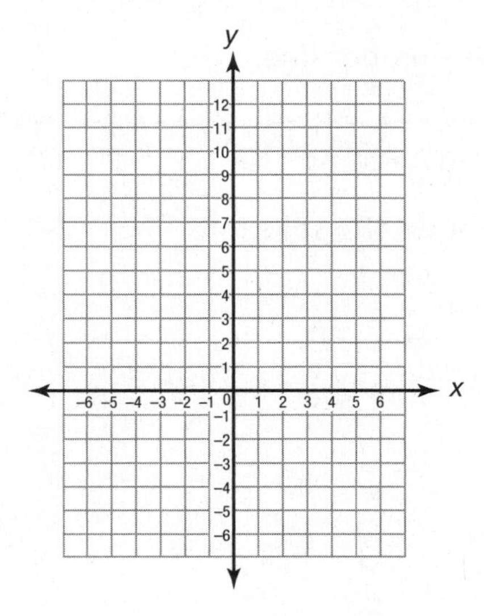

24 Consider this inequality.

$4x - 9 \geq -5$

Part A

Solve the inequality.

Answer ______________

Part B

Graph the solution on this number line.

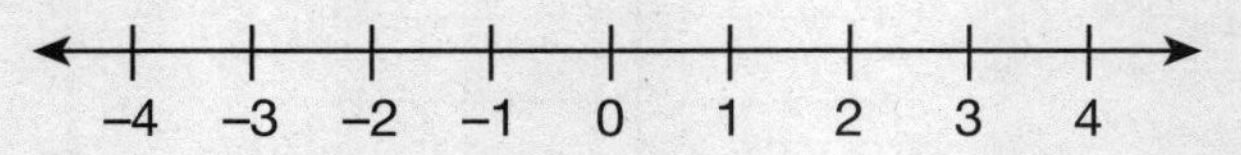

Explain why your graph shows the solution.

____

STRAND

3 Geometry

** Grade 8 May—June Indicators

18 Angle Pairs

8.G.1, 8.G.2, 8.G.3, 8.G.6, 8.A.12

Getting the Idea

Angles are measured in degrees (°). There are some special kinds of angles you should know.

Types of Angles	Diagram
Adjacent angles have a common side and do not overlap. They have no interior points in common.	$\angle WXY$ and $\angle YXZ$ are adjacent angles.
Vertical angles are two nonadjacent angles formed by intersecting lines. Vertical angles are congruent, or equal in measure.	$\angle ABD \cong \angle CBE$ $\angle ABC \cong \angle DBE$
Complementary angles are two angles whose measures have a sum of 90°.	$m\angle FGH + m\angle JKL = 70° + 20° = 90°$ ($m\angle FGH$ means "the measure of $\angle FGH$.").
Supplementary angles are two angles whose measures have a sum of 180°.	$m\angle PQR + m\angle RQS = 65° + 115° = 180°$

EXAMPLE 1

The diagram below shows $\overline{AD}$, $\overline{EB}$, and $\overline{EC}$.

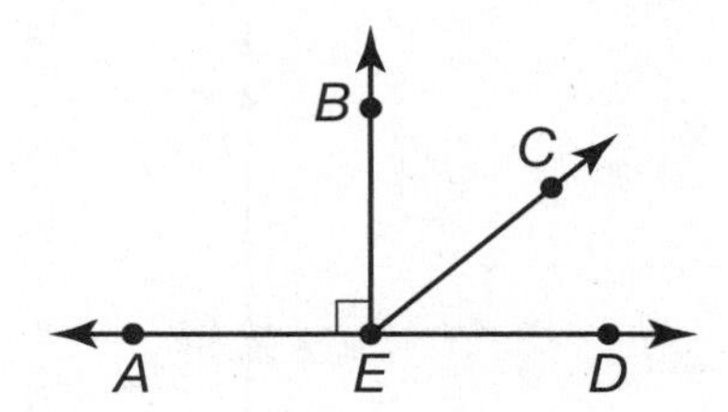

Are $\angle BEC$ and $\angle CED$ complementary angles, supplementary angles, or neither?

STRATEGY **Use the properties of right angles and straight angles to classify the angle pair.**

STEP 1 Identify any right angles and straight angles.

Remember that a right angle measures 90°, and a straight angle measures 180°.

$\angle AED$ is a straight angle because AD is a line, so $\angle AED$ measures 180°.

$\angle AEB$ is marked as a right angle, so $\angle AEB$ measures 90°.

STEP 2 Use the information from Step 1 to identify the relationship between $\angle BEC$ and $\angle CED$.

Write an equation.

$m\angle BED + m\angle AEB = m\angle AED$

Substitute the known angle measures to find $m\angle BED$.

$m\angle BED + 90 = 180$

$m\angle BED = 90$

$m\angle BEC + m\angle CED = m\angle BED = 90$

The sum of the measures of the two angles is 90°.

SOLUTION **$\angle BEC$ and $\angle CED$ are complementary angles.**

EXAMPLE 2

Angles *LOM* and *MON* are complementary angles. The measure of $\angle LOM$ is 50°.

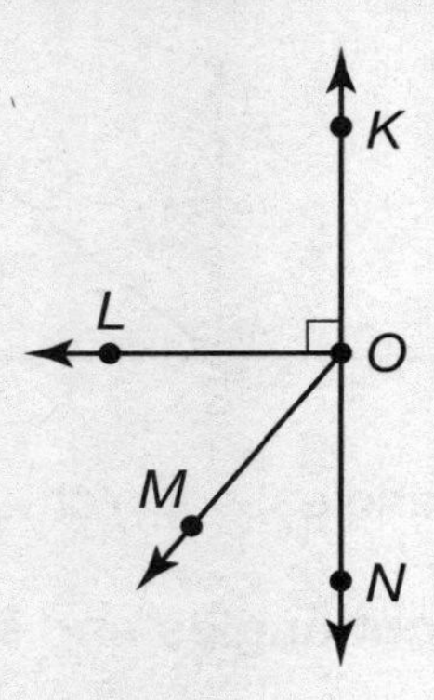

What is the measure of $\angle MON$?

STRATEGY **Use the information in the diagram and the meaning of complementary angles.**

Use the meaning of complementary angles to write an equation.

The measures of complementary angles sum to 90°.

$m\angle MON + m\angle LOM = 90$

$\angle LOM = 50$

$m\angle MON + 50 = 90$

$90 - 50 = \angle MON$

$m\angle MON = 40$

SOLUTION **The measure of $\angle MON$ is 40°.**

EXAMPLE 3

In the diagram below, $\angle ABD$ measures 20°.

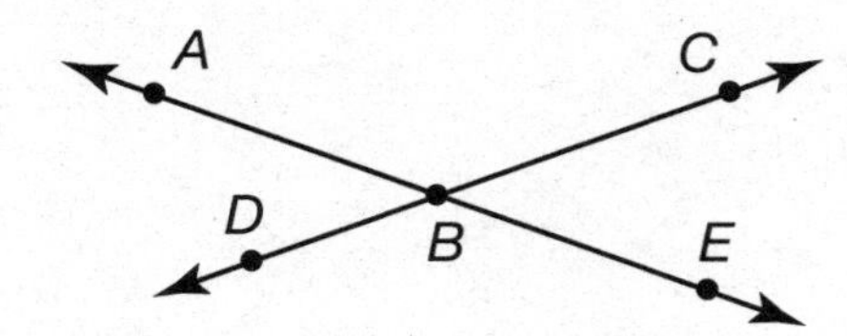

What are the measures of $\angle CBE$, $\angle ABC$, and $\angle DBE$?

STRATEGY **Use the properties of vertical and supplementary angles.**

STEP 1 Use the properties of supplementary angles.

$\angle ABD$ and $\angle DBE$ form a straight angle, which has a measure of 180°. So, $\angle ABD$ and $\angle DBE$ are supplementary angles.

$\angle ABD$ measures 20°, so the measure of $\angle DBE = 180° - 20° = 160°$.

STEP 2 Identify the vertical angles.

$\angle ABD$ and $\angle CBE$ are vertical angles.

$\angle DBE$ and $\angle ABC$ are vertical angles.

STEP 3 Use the properties of vertical angles to find the other two angle measures.

Vertical angles are congruent.

$m\angle ABD = m\angle CBE = 20°$

$m\angle DBE = m\angle ABC = 160°$

SOLUTION **$\angle CBE$ measures 20°; $\angle ABC$ measures 160°; and $\angle DBE$ measures 160°.**

COACHED EXAMPLE

The measure of $\angle PQR$ is 50°.

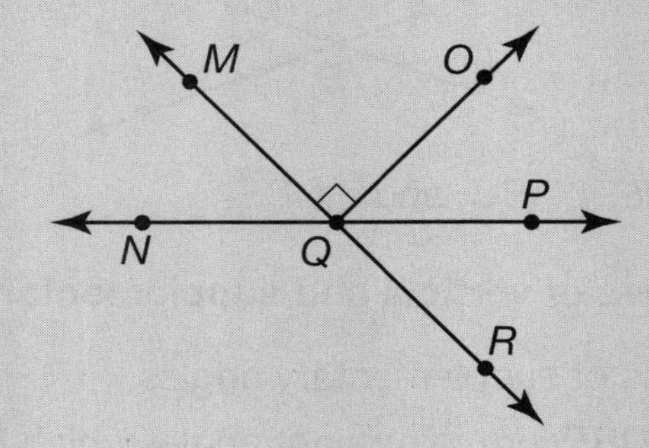

What is $m\angle NQR$?

THINKING IT THROUGH

$\angle PQR$ and $\angle NQR$ are ________________ angles.

So, $m\angle PQR + m\angle NQR =$ _______.

$m\angle NQR =$ ______ − ________ = __________.

The measure of $\angle NQR$ is ___________.

Lesson Practice

Choose the correct answer.

Use this diagram for questions 1 and 2.

Lines *MN* and *PQ* intersect at point *R*.

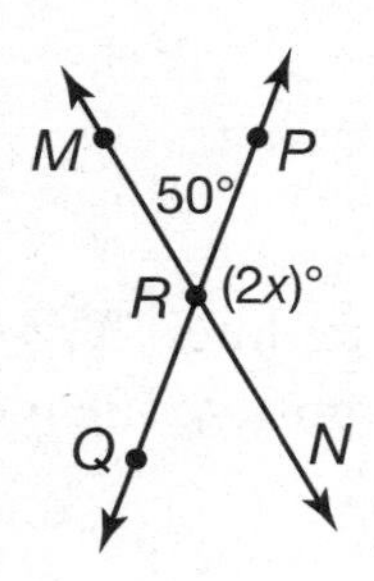

1. What is the measure of $\angle QRN$?

   A. 30°
   B. 40°
   C. 50°
   D. 130°

2. What is the value of $\angle PRN$?

   A. 25
   B. 65
   C. 130
   D. 180

Use this diagram for questions 3 and 4.

Ray *BE* is perpendicular to line *AC*.

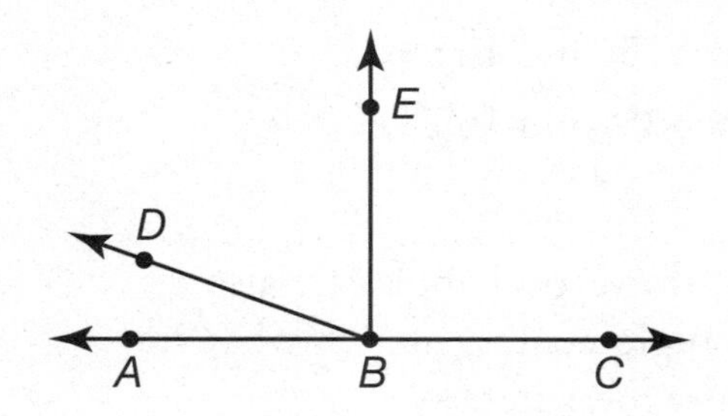

3. If $m\angle ABD = 25°$, what is $m\angle DBE$?

   A. 25°
   B. 55°
   C. 65°
   D. 70°

4. If $m\angle ABD = 25°$, what is $m\angle DBC$?

   A. 25°
   B. 65°
   C. 90°
   D. 155°

5. Which of the following could be the measures of a pair of complementary angles?

   A. 8° and 82°
   B. 18° and 82°
   C. 18° and 162°
   D. 172° and 172°

6. Which of the following could be the measures of a pair of supplementary angles?

A. 15° and 15°
B. 25° and 65°
C. 15° and 155°
D. 25° and 155°

7. If two vertical angles are also complementary angles, what are their measures?

A. 30° and 60°
B. 45° and 45°
C. 80° and 100°
D. 90° and 90°

8. Which types of angles are always congruent?

A. vertical
B. complementary
C. supplementary
D. exterior

9. If an angle measures 56°, what is the measure of its complementary angle?

A. 34°
B. 44°
C. 56°
D. 144°

10. What is the value of *b*?

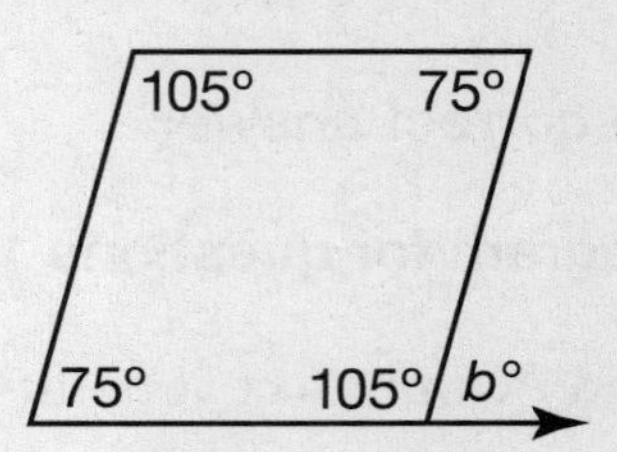

Answer ________

11. In this diagram, lines *JK*, *MN*, and *PQ* intersect at point *O*. The measure of ∠*NOK* is 55°.

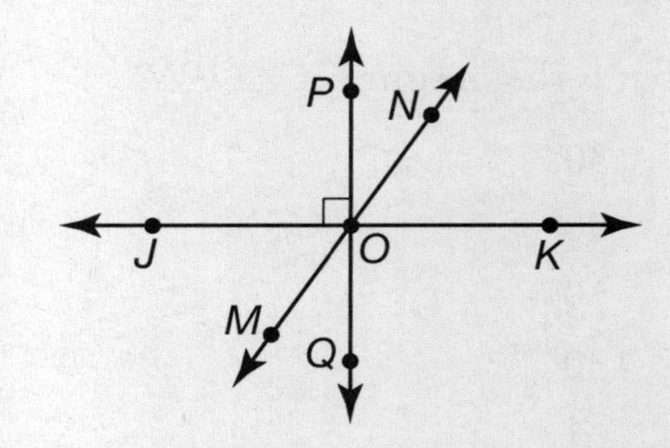

What is the measure of ∠*POM*?

Answer ____________

19 Parallel Lines and Transversals

8.G.4, 8.G.5, 8.A.12

Getting the Idea

Parallel lines are lines in the same plane that remain the same distance apart and never meet. A **transversal** is a line that intersects two or more other lines.

When a transversal intersects parallel lines, some special angle pairs are formed. In the diagram below, $\overleftrightarrow{AB}$ and $\overleftrightarrow{CD}$ are parallel, and $\overleftrightarrow{RT}$ is a transversal.

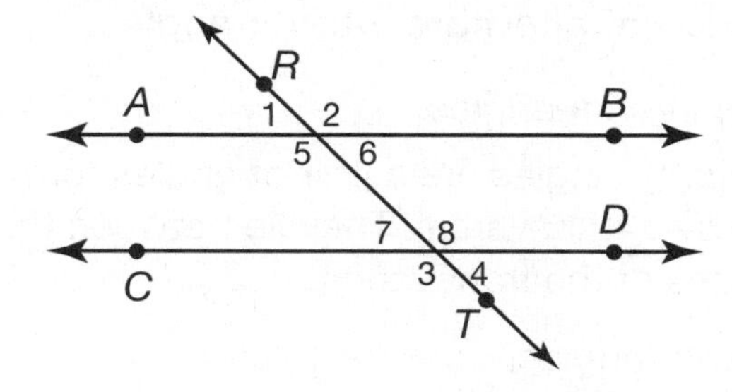

Interior angles lie between $\overleftrightarrow{AB}$ and $\overleftrightarrow{CD}$.

$\angle 5$, $\angle 6$, $\angle 7$, and $\angle 8$ are interior angles.

Exterior angles lie outside $\overleftrightarrow{AB}$ and $\overleftrightarrow{CD}$.

$\angle 1$, $\angle 2$, $\angle 3$, and $\angle 4$ are exterior angles.

Here are some special angle pairs.

Alternate interior angles lie inside the parallel lines and are on opposite sides of the transversal. Alternate interior angles are congruent.

$\angle 5 \cong \angle 8 \quad \angle 6 \cong \angle 7$

Alternate exterior angles lie outside the parallel lines and are on opposite sides of the transversal. Alternate exterior angles are congruent.

$\angle 1 \cong \angle 4 \quad \angle 2 \cong \angle 3$

In a pair of **corresponding angles**, one angle lies outside the parallel lines and one angle lies inside the parallel lines. Both are on the same side of the transversal. Corresponding angles are congruent.

$\angle 1 \cong \angle 7 \quad \angle 5 \cong \angle 3$

$\angle 2 \cong \angle 8 \quad \angle 6 \cong \angle 4$

EXAMPLE 1

Lines *a* and *b* are parallel, and line *c* is a transversal. What are two pairs of alternate interior angles?

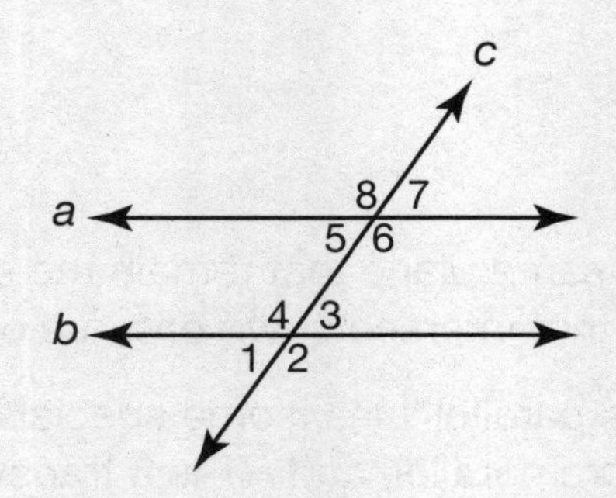

STRATEGY **Use the definition of alternate interior angles.**

STEP 1 Define alternate interior angles.

Alternate interior angles are a pair of angles formed by parallel lines intersected by a transversal. They lie between the parallel lines and are on opposite sides of the transversal.

STEP 2 Identify the interior angles.

$\angle 3$, $\angle 4$, $\angle 5$, and $\angle 6$ are interior angles.

STEP 3 Identify the two pairs of alternate interior angles.

Which pairs of angles are on opposite sides of the transversal?

$\angle 3$ and $\angle 5$

$\angle 4$ and $\angle 6$

SOLUTION **The two pairs of alternate interior angles are $\angle 3$ and $\angle 5$, and $\angle 4$ and $\angle 6$.**

When you know the measure of one angle formed by parallel lines that are cut by a transversal, the properties of angle pairs allow you to find the measures of other angles.

EXAMPLE 2

Lines *a* and *b* are parallel, and line *c* is a transversal. If $m\angle 7 = 65°$, what is $m\angle 2$?

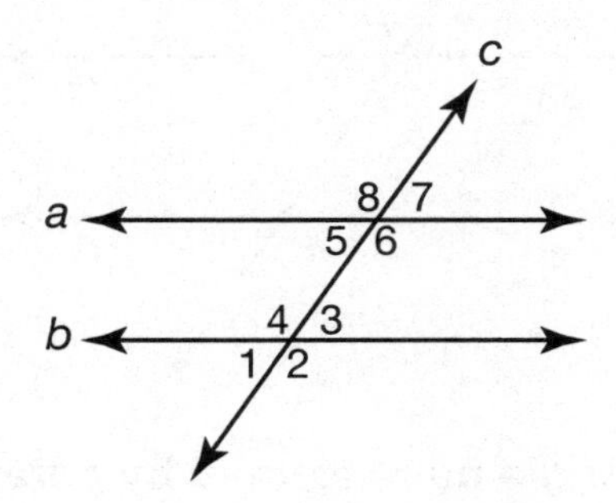

STRATEGY **Use the properties of angle pairs formed by parallel lines and a transversal.**

STEP 1 $\angle 7$ is one angle in a pair of alternate exterior angles. What is the other angle in the pair?

The other angle is $\angle 1$.

STEP 2 What is the measure of $\angle 1$?

$\angle 1 \cong \angle 7$, so $m\angle 1 = m\angle 7 = 65°$

STEP 3 What is the relationship between $\angle 1$ and $\angle 2$?

$\angle 1$ and $\angle 2$ are supplementary angles.

STEP 4 Use the properties of supplementary angles.

The sum of the measures of supplementary angles is 180°.

$m\angle 1 + m\angle 2 = 180$

$65 + m\angle 2 = 180$

$m\angle 2 = 180 - 65$

$m\angle 2 = 115$

SOLUTION **The measure of $\angle 2$ is 115°.**

EXAMPLE 3

$\overleftrightarrow{AC}$ and $\overleftrightarrow{EG}$ are parallel, and $\overleftrightarrow{DH}$ is a transversal. The measure of $\angle CBD$ is 125°. What is the measure of $\angle EFB$?

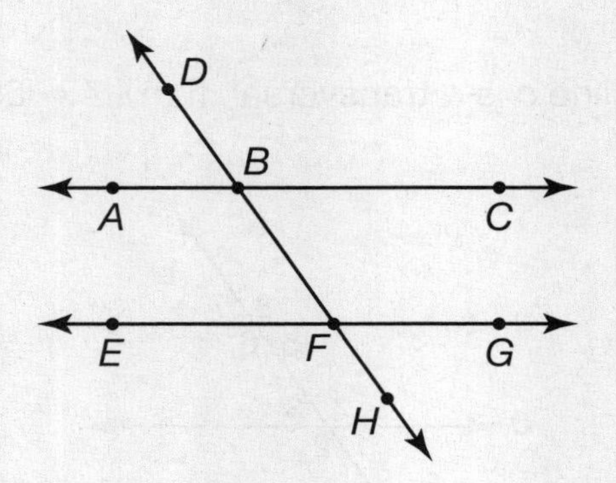

STRATEGY **Use properties of angle pairs formed by parallel lines and a transversal.**

STEP 1 Find an angle that is congruent to $\angle CBD$.

$\angle GFB$ and $\angle CBD$ are corresponding angles.

$\angle GFB \cong \angle CBD$

STEP 2 What is the measure of $\angle GFB$?

$m\angle GFB = m\angle CBD = 125°$

STEP 3 What is the relationship between $\angle GFB$ and $\angle EFB$?

$\angle GFB$ and $\angle EFB$ are supplementary angles.

STEP 4 Use the relationship to find $m\angle EFB$.

$m\angle GFB + m\angle EFB = 180$

$125 + m\angle EFB = 180$

$m\angle EFB = 180 - 125$

$m\angle EFB = 55$

SOLUTION **The measure of $\angle EFB$ is 55°.**

Note: There are other ways to find $m\angle EFB$ in Example 3. For example, you can also use the fact that $\angle CBD$ and $\angle EFH$ are congruent alternate exterior angles, and that $\angle EFB$ and $\angle EFH$ are supplementary angles.

COACHED EXAMPLE

Lines *d* and *e* are parallel, and line *f* is a transversal. If m∠4 = 150°, what is m∠5?

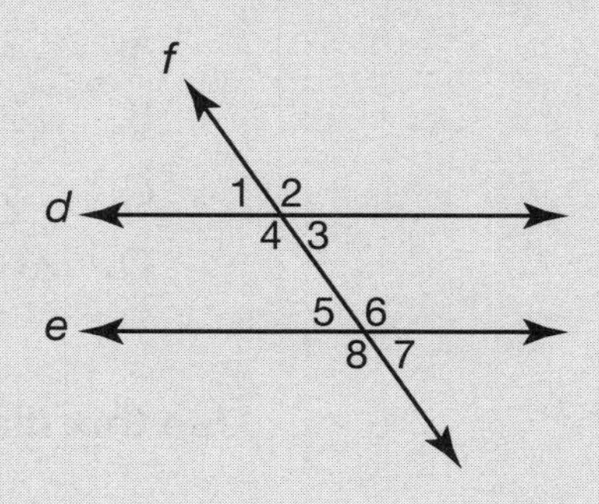

THINKING IT THROUGH

Is ∠4 an interior or exterior angle? ________________

∠4 and ∠6 are ______________ ______________ angles.

So, ∠4 is ______________ to ∠6.

The measure of ∠4 is 150°, so ∠6 measures ________.

∠5 and ∠6 are ________________ angles.

So, the sum of the measures of ∠5 and ∠6 is __________.

Find the measure of ∠5.

_________ – _________ = ______

The measure of ∠5 is _______.

Lesson Practice

Choose the correct answer.

Use this diagram for questions 1–3.

Lines *l* and *m* are parallel, and line *t* is a transversal.

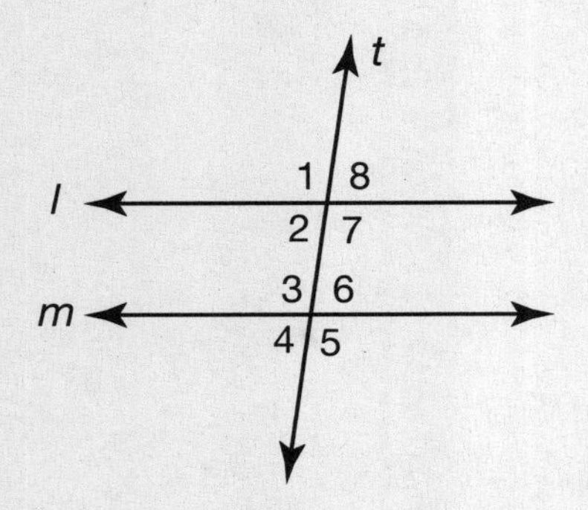

1. Which pair of angles are alternate interior angles?

   A. $\angle 1$ and $\angle 6$
   B. $\angle 2$ and $\angle 3$
   C. $\angle 3$ and $\angle 7$
   D. $\angle 4$ and $\angle 8$

2. Which pair of angles are alternate exterior angles?

   A. $\angle 1$ and $\angle 4$
   B. $\angle 3$ and $\angle 8$
   C. $\angle 4$ and $\angle 8$
   D. $\angle 5$ and $\angle 8$

3. Which pair of angles are corresponding angles?

   A. $\angle 1$ and $\angle 4$
   B. $\angle 2$ and $\angle 3$
   C. $\angle 5$ and $\angle 8$
   D. $\angle 6$ and $\angle 8$

Use this diagram for questions 4 and 5.

Lines *r* and *s* are parallel, and line *t* is a transversal.

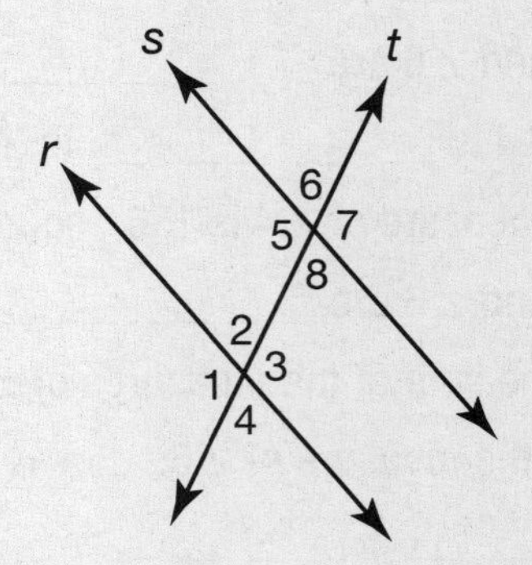

4. Which angle is congruent to $\angle 3$?

   A. $\angle 4$ **C.** $\angle 6$
   B. $\angle 5$ **D.** $\angle 8$

5. If $\angle 1$ measures 115°, what is the measure of $\angle 8$?

   A. 115° **C.** 75°
   B. 85° **D.** 65°

6. In this diagram, lines e and f are parallel, and line g is a transversal. If $\angle 3$ measures 66°, what is the measure of $\angle 6$?

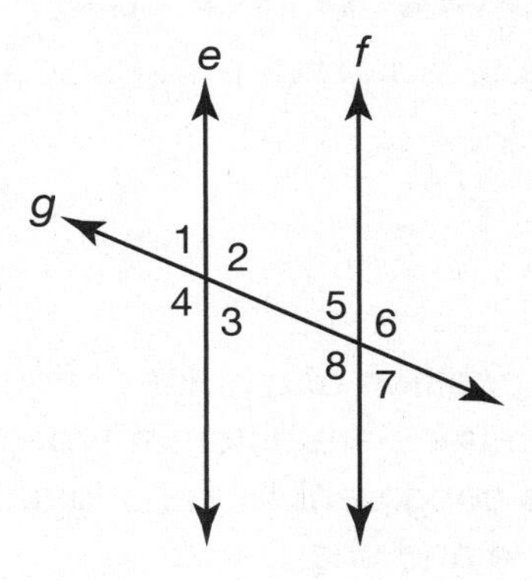

Answer __________

EXTENDED-RESPONSE QUESTION

7. In the diagram below, lines a and b are parallel, and line c is a transversal. $\angle 6$ measures $(3x - 20)°$ and $\angle 2$ measures $(2x + 10)°$.

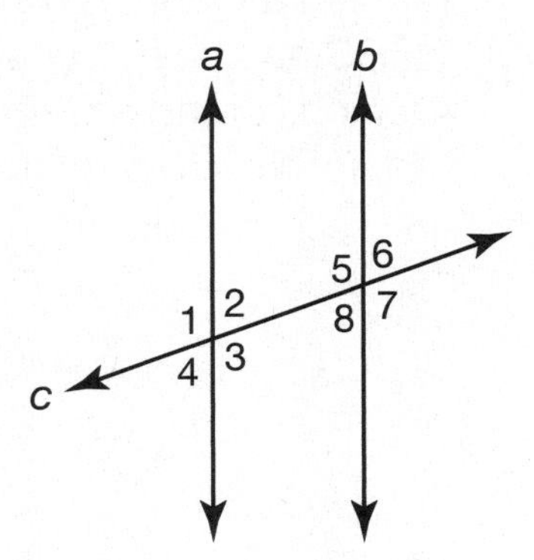

Part A

Write an equation to solve for x. What is the value of x? Explain how you determined your answer.

Answer __________

____

____

Part B

Use your answer to Part A to find the measure of $\angle 4$. Explain how you determined your answer.

Answer __________

____

____

Lesson

20 Translations, Rotations, and Reflections

8.G.7, 8.G.8, 8.G.9, 8.G.10, 8.G.12

Getting the Idea

A **transformation** changes the position of a point or figure on a coordinate plane. Translations, reflections, and rotations are different types of transformations. The transformed figure, or **image**, is congruent to the original figure. So, the image and the original figure have the same size and shape.

You can name the image by using one or more prime symbols (′). If figure *ABC* is transformed, its image can be called *A′B′C′*.

In a **translation**, every point in a figure moves the same distance in one or two directions within a plane. A translation is also called a slide. On the coordinate plane below, triangle *ABC* has been translated 6 units right and 2 units down. The image is triangle *A′B′C′*, which is read as "triangle A prime B prime C prime." Notice that in a translation, the image is not turned or flipped in any way.

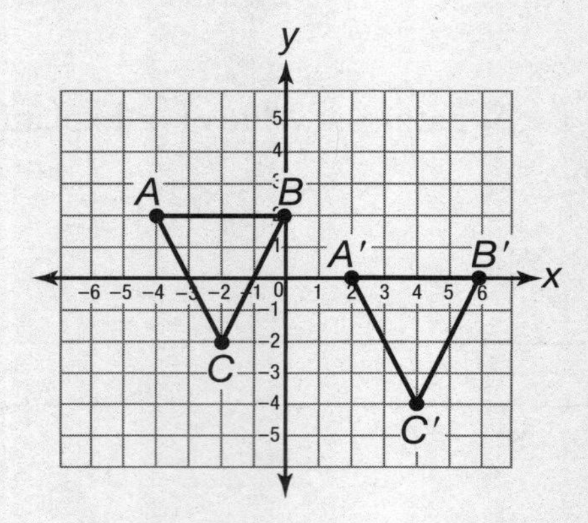

A **rotation** turns a figure around a point. Rotations are also called turns. Some examples of turns are a 90° turn (a quarter turn) and a 180° turn (a half turn).

On the coordinate plane below, trapezoid *WXYZ* has been rotated 90° counterclockwise about the origin to form the trapezoid *W′X′Y′Z′*.

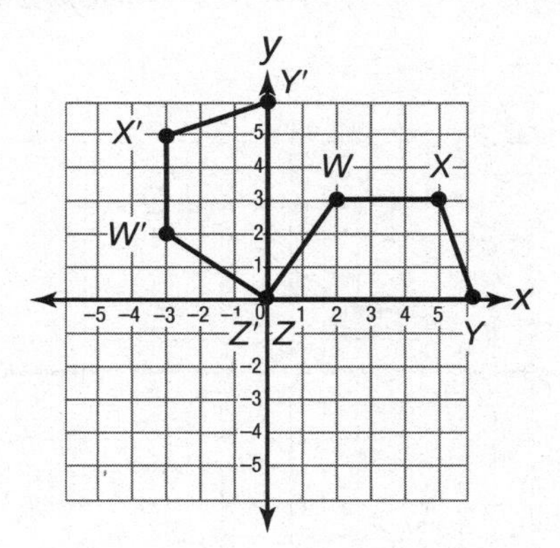

For rotations about the origin, if a point in a figure has coordinates (x, y), then:

- the coordinates of that point's image after a 90° counterclockwise rotation are $(-y, x)$
- the coordinates of that point's image after a 90° clockwise rotation are $(y, -x)$
- the coordinates of that point's image after a 180° rotation are $(-x, -y)$.

Note: A 270° clockwise rotation is the same as a 90° counterclockwise rotation. Similarly, a 270° counterclockwise rotation is the same as a 90° clockwise rotation.

A **reflection** produces a mirror image of a figure by flipping it over a line. That is why reflections are also called flips. On the coordinate plane below, triangle *DEF* has been reflected over the *x*-axis to form triangle *D′E′F′*.

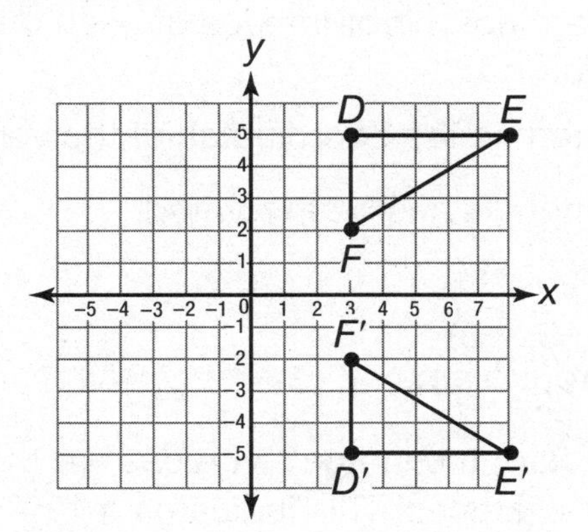

When a figure is reflected over the *x*-axis:

- the *x*-coordinates in the original figure and the image are the same
- the *y*-coordinates of the image are the opposite of those in the original figure.

When a figure is reflected over the *y*-axis:

- the *y*-coordinates in the original figure and the image are the same
- the *x*-coordinates of the image are the opposite of those in the original figure.

EXAMPLE 1

Triangle PQR and its image, triangle $P'Q'R'$, are shown below.

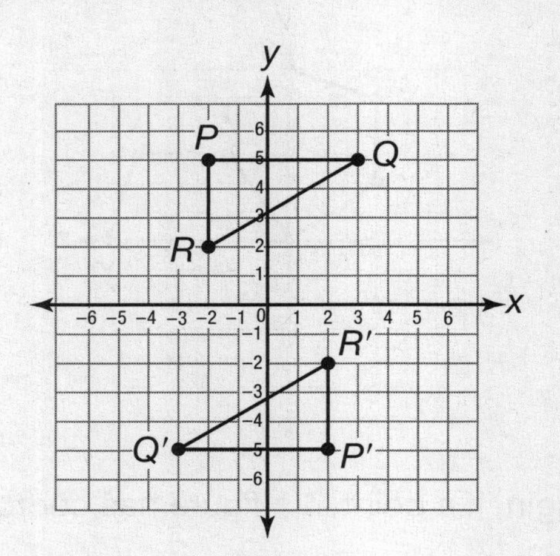

What transformation was applied to triangle PQR to get triangle $P'Q'R'$?

STRATEGY **Compare the triangles with the meanings of translation, rotation, and reflection.**

STEP 1 Determine if the triangle has been translated.

The triangles do not have the same orientation, so the transformation is not a translation.

STEP 2 Determine if the triangle has been reflected.

The triangles are not mirror images of each other, so the transformation is not a reflection.

There is a pattern in the coordinates of the vertices of the figures:

STEP 3 Determine if the triangle has been rotated.

$P(-2, 5) \rightarrow P'(2, -5)$

$Q(3, 5) \rightarrow Q'(-3, -5)$

$R(-2, 2) \rightarrow R'(2, -2)$

The coordinates of the image's vertices are the opposites of those of the original figure's vertices. This indicates a 180° rotation about the origin.

SOLUTION **Triangle *PQR* was rotated 180° to get triangle *P′Q′R′*.**

EXAMPLE 2

Translate quadrilateral *JKLM* 5 units to the left and 3 units up.

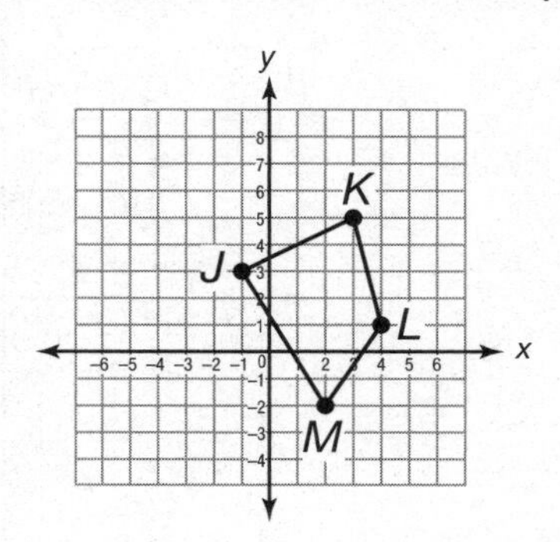

STRATEGY **Translate each vertex 5 units to the left and 3 units up.**

STEP 1 Identify the coordinates of each vertex.

The coordinates of J are $(-1, 3)$.

The coordinates of K are $(3, 5)$.

The coordinates of L are $(4, 1)$.

The coordinates of M are $(2, -2)$.

STEP 2 Translate each vertex.

To translate a point 5 units to the left, subtract 5 from its x-coordinate.

To translate a point 3 units up, add 3 to its y-coordinate.

$J(-1, 3) \rightarrow J'(-1 - 5, 3 + 3) \rightarrow J'(-6, 6)$

$K(3, 5) \rightarrow K'(3 - 5, 5 + 3) \rightarrow K'(-2, 8)$

$L(4, 1) \rightarrow L'(4 - 5, 1 + 3) \rightarrow L'(-1, 4)$

$M(2, -2) \rightarrow M'(2 - 5, -2 + 3) \rightarrow M'(-3, 1)$

STEP 3 Graph the image.

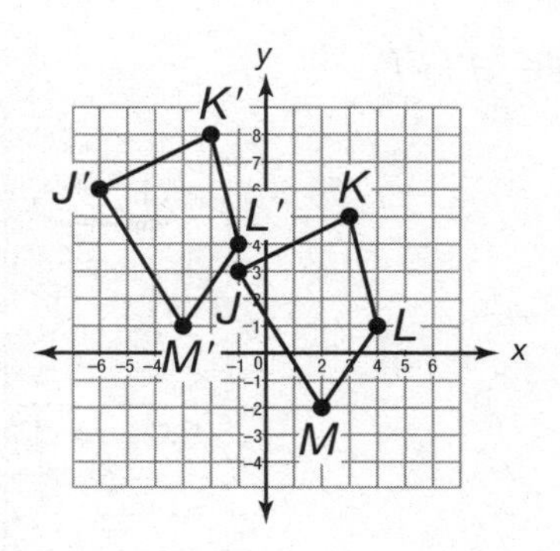

SOLUTION **Step 3 shows the translation of quadrilateral *JKLM* 5 units to the left and 3 units up.**

EXAMPLE 3

Reflect triangle *GHI* over the *y*-axis.

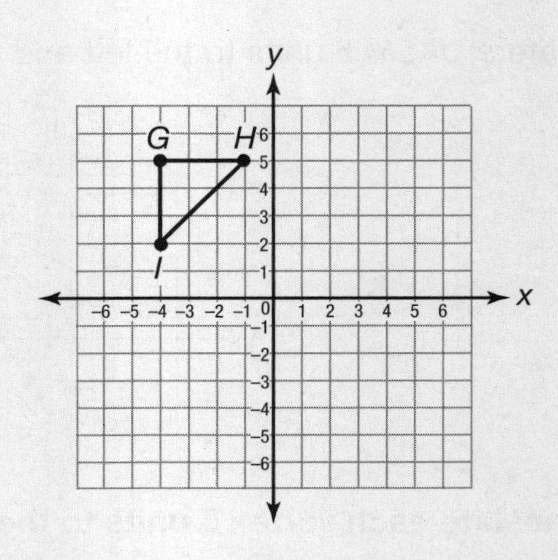

STRATEGY **Write the ordered pairs for vertices *G*, *H*, and *I*. Then use the rule for the effects of a reflection over the *y*-axis to determine the coordinates of *G′*, *H′*, and *I′*.**

STEP 1 Identify the coordinates of the vertices of triangle *GHI*.

The coordinates of *G* are $(-4, 5)$.

The coordinates of *H* are $(-1, 5)$.

The coordinates of *I* are $(-4, 2)$.

STEP 2 Identify the coordinates of the vertices of image triangle *G′H′I′*.

Recall that when a figure is reflected over the *y*-axis, the *y*-coordinates do not change, but the *x*-coordinates of the image are the opposite of those of the original figure.

$G(-4, 5) \rightarrow G'(4, 5)$

$H(-1, 5) \rightarrow H'(1, 5)$

$I(-4, 2) \rightarrow I'(4, 2)$

STEP 3 Plot and draw triangle *G′H′I′*.

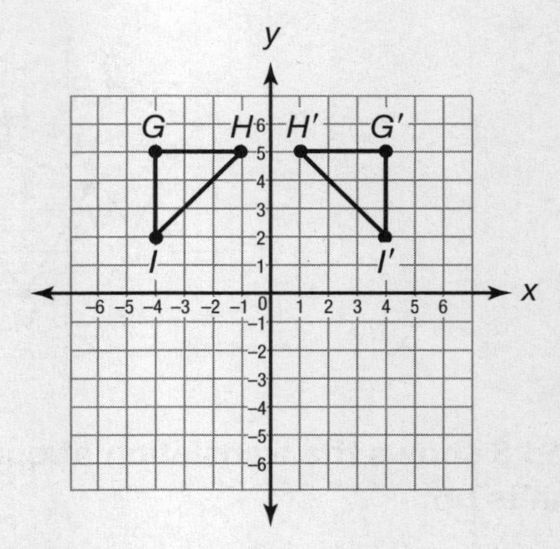

SOLUTION **Step 3 shows the reflection of triangle *GHI* over the *y*-axis.**

EXAMPLE 4

Rectangle *QRST* is rotated 180° about the origin. Find the coordinates of *Q′R′S′T′* and graph the image.

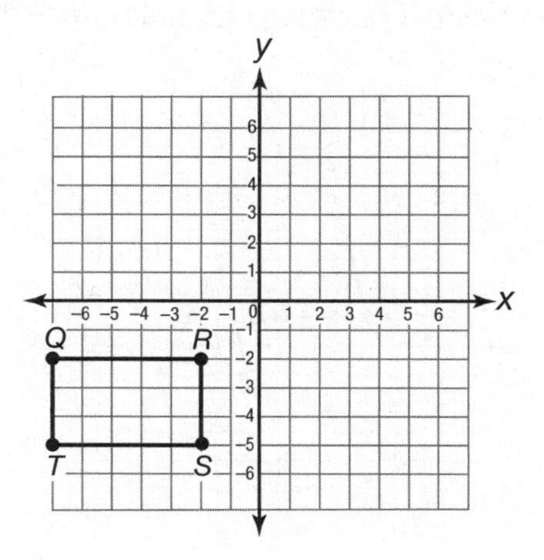

STRATEGY **Rotate each vertex 180° about the origin.**

STEP 1 Identify the coordinates of each vertex of the image.

When a figure is rotated 180° about the origin, the signs of the *x*- and *y*-coordinates change.

The coordinates of *Q* are (−7, −2), so the coordinates of *Q′* are (7, 2).

The coordinates of *R* are (−2, −2), so the coordinates of *R′* are (2, 2).

The coordinates of *S* are (−2, −5), so the coordinates of *S′* are (2, 5).

The coordinates of *T* are (−7, −5), so the coordinates of *T′* are (7, 5).

STEP 2 Plot rectangle *Q′R′S′T′*.

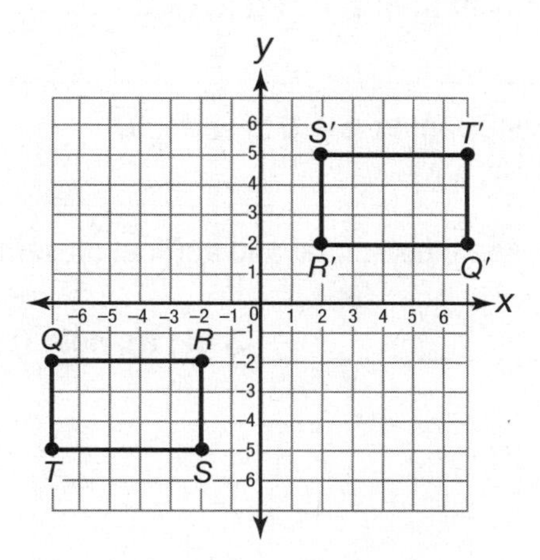

SOLUTION **The coordinates of rectangle *Q′R′S′T′* are *Q′*(7, 2), *R′*(2, 2), *S′*(2, 5), and *T*(7, 5). The graph of *Q′R′S′T′* is shown in Step 2.**

COACHED EXAMPLE

What kind of transformation created the image triangle *Q′R′S′*?

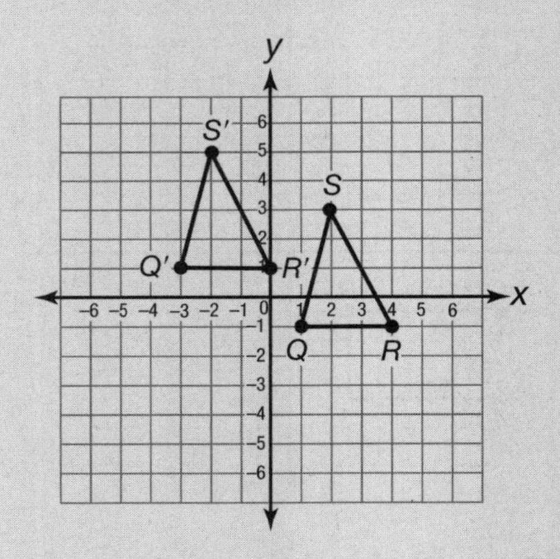

THINKING IT THROUGH

Are the triangles congruent? __________

Is triangle *Q′R′S′* a reflection of triangle *QRS* over the *x*-axis?

____

Is triangle *Q′R′S′* a reflection of triangle *QRS* over the *y*-axis?

____

Is triangle *Q′R′S′* a rotation of triangle *QRS* about the origin?

____

Describe the horizontal movement from vertex *Q* to vertex *Q′*.

____

Describe the vertical movement from vertex *Q* to vertex *Q′*.

____

Do vertices *R* and *S* have the same horizontal and vertical movement as vertex *Q*? ____________

Triangle *QRS* was ________________________ to get triangle *Q′R′S′*.

Lesson Practice

Choose the correct answer.

1. Which kind of transformation was applied to triangle *NOP* to get triangle *N′O′P′*?

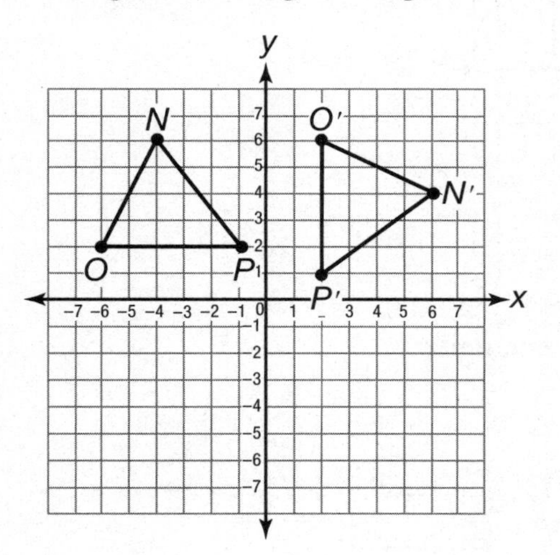

A. a reflection over the y-axis
B. a 90° clockwise rotation about the origin
C. a 90° counterclockwise rotation about the origin
D. a 180° rotation about the origin

2. Point $C(3, -8)$ is reflected over the x-axis. What are the coordinates of its image, C'?

A. $(3, 8)$ **C.** $(-8, 3)$
B. $(-3, 8)$ **D.** $(8, -3)$

3. Point $J(6, 4)$ is rotated 180° about the origin. What are the coordinates of its image, J'?

A. $(-6, -4)$ **C.** $(6, -4)$
B. $(-6, 4)$ **D.** $(-4, -6)$

4. Which kind of transformation was applied to triangle *UVW* to get triangle *U′V′W′*?

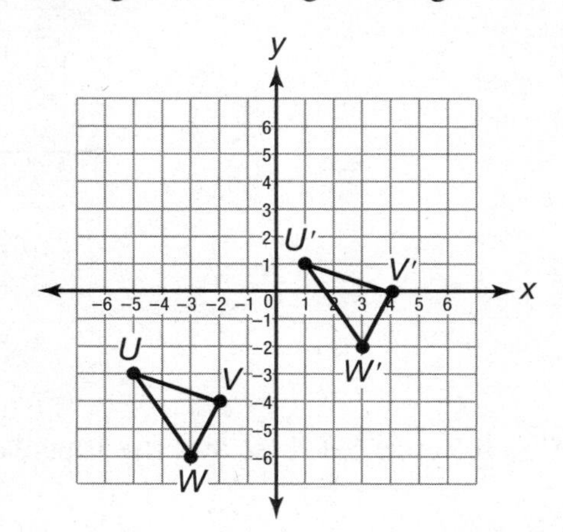

A. a reflection over the x-axis
B. a reflection over the y-axis
C. a rotation
D. a translation

5. Which kind of transformation was applied to *ABCD* to produce rectangle *A′B′C′D′*?

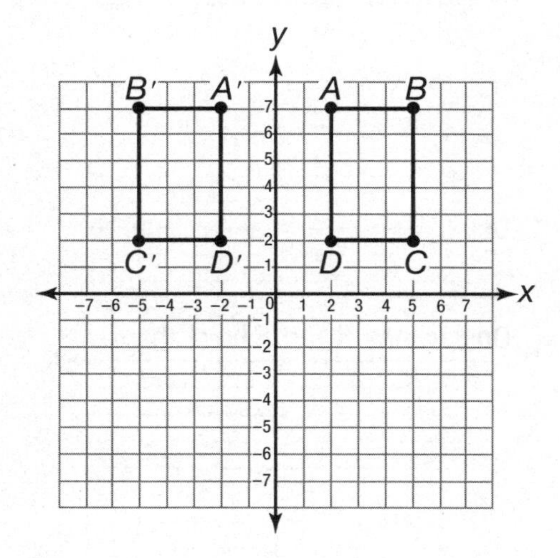

A. a reflection over the x-axis
B. a reflection over the y-axis
C. a rotation
D. a translation

6. The rectangle below will be rotated 90° counterclockwise about the origin.

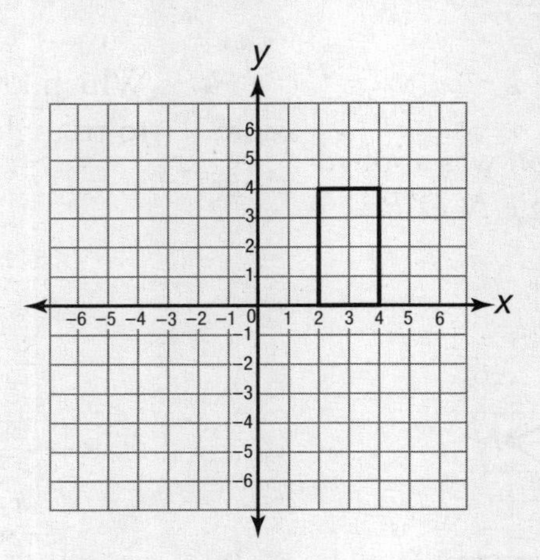

Which figure below shows the image of the original rectangle?

A.

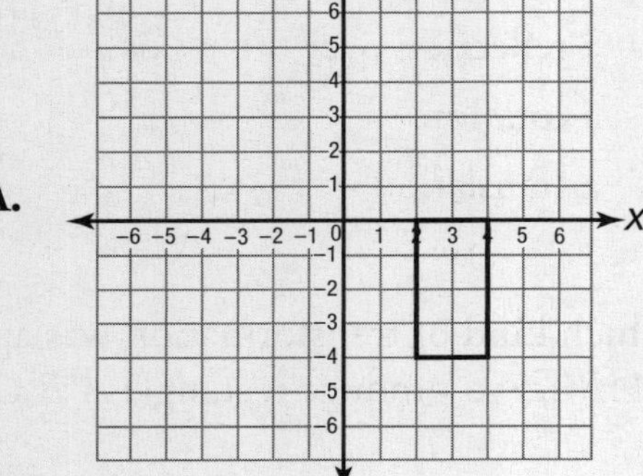

B.

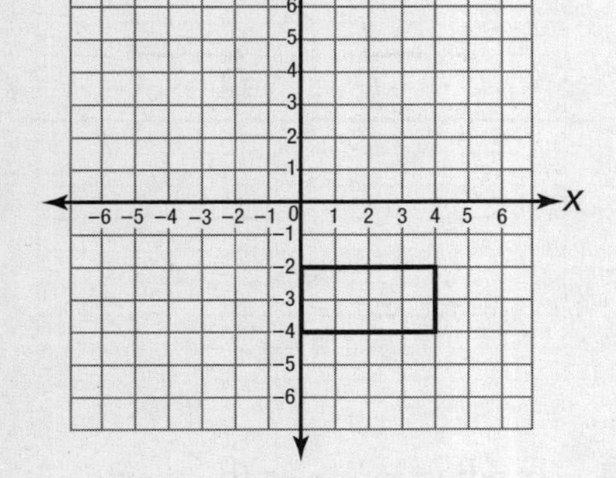

C.

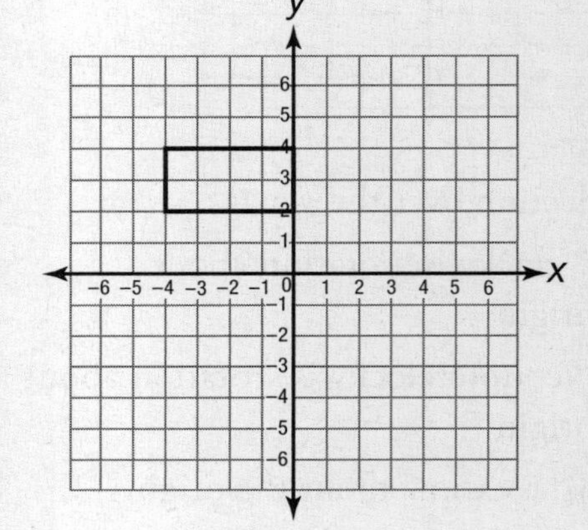

D.

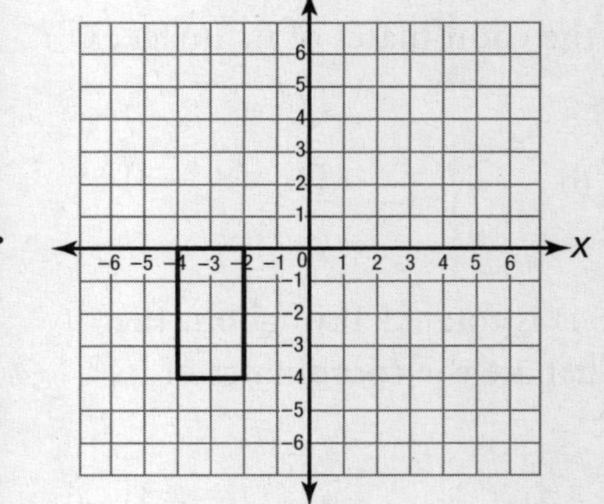

7. On the coordinate plane below, graph triangle $R'S'T'$, the image of triangle RST after a translation 6 units to the right and 4 units down. What are the coordinates of T'?

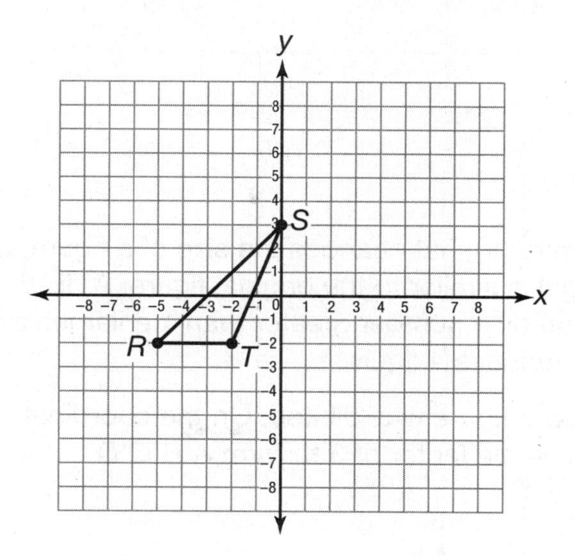

Answer __________

EXTENDED-RESPONSE QUESTION

8. Triangles DEF and $D'E'F'$ are plotted below.

Part A What kind of transformation was applied to triangle DEF to get $D'E'F'$?

Answer __________________

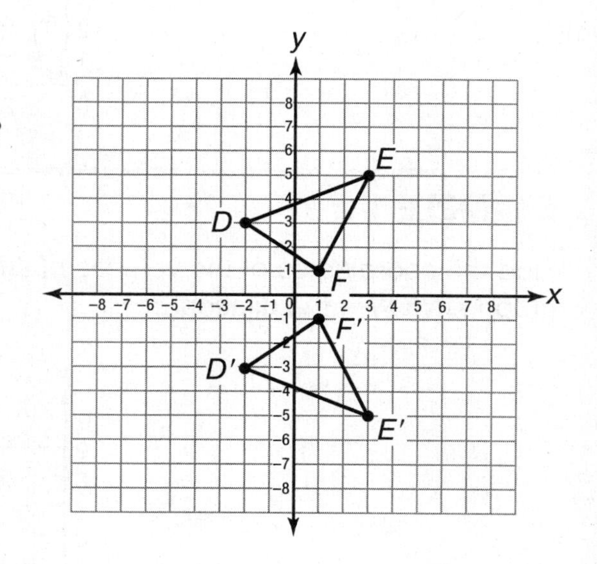

Part B Describe how the coordinates of point D changed to the coordinates of D'.

____

____

21 Dilations

8.G.7, 8.G.11, 8.G.12

Getting the Idea

A **dilation** is a transformation that changes the size of a figure, but not its shape. The dilated figure (the image) is similar to the original figure. A dilation may be an enlargement or a reduction. A dilation by a number greater than 1 enlarges a figure. A dilation by a number less than as 1 reduces a figure.

A **scale factor** tells how a figure was dilated. On the coordinate plane below, rectangle *ABCD* was dilated by a scale factor of 3 to form *A′B′C′D′*.

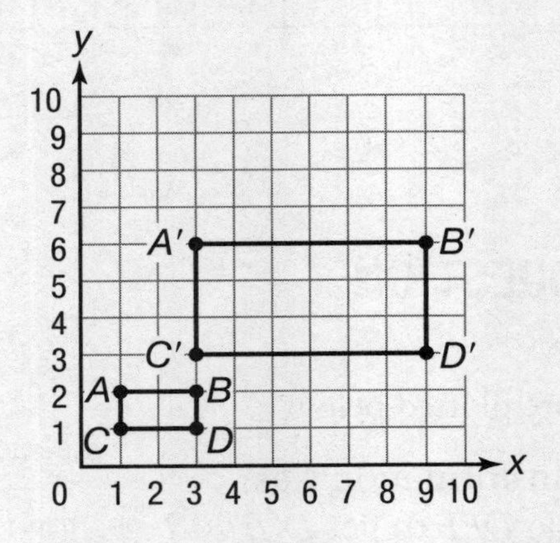

EXAMPLE 1

Find the coordinates of the vertices of △*NOP* after a dilation with a scale factor of 2. Then graph △*N′O′P′*, the dilation.

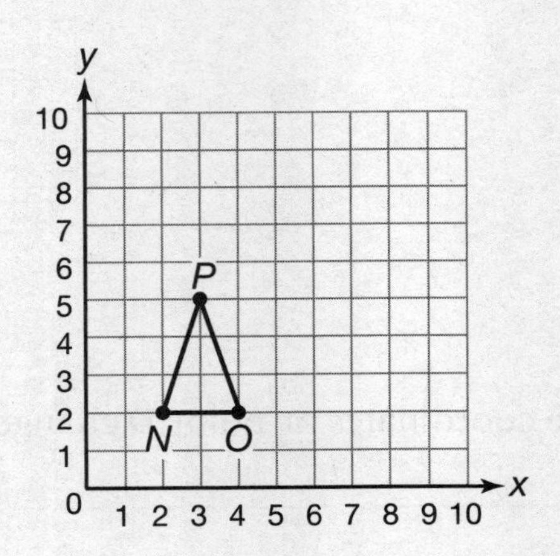

STRATEGY **Use the scale factor to identify the coordinates of the vertices of the image.**

STEP 1 Use the scale factor.

The scale factor is 2, so multiply each coordinate of each vertex by 2.

$N(2, 2) \rightarrow N'(2 \times \mathbf{2}, 2 \times \mathbf{2}) \rightarrow N'(4, 4)$

$O(4, 2) \rightarrow O'(4 \times \mathbf{2}, 2 \times \mathbf{2}) \rightarrow O'(8, 4)$

$P(3, 5) \rightarrow P'(3 \times \mathbf{2}, 5 \times \mathbf{2}) \rightarrow P'(6, 10)$

STEP 2 Graph $\triangle N'O'P'$.

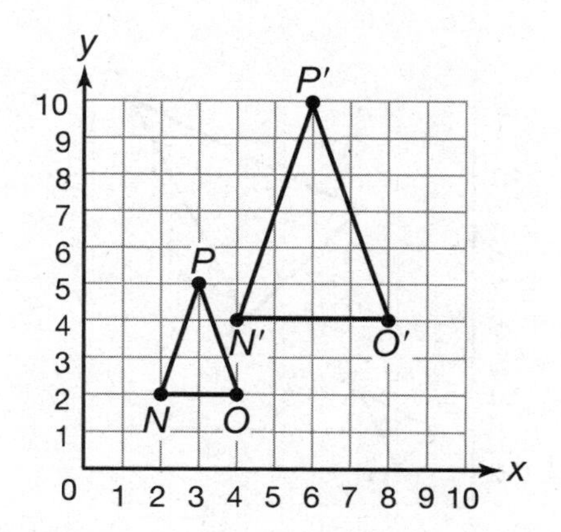

SOLUTION **The coordinates of the vertices of $\triangle NOP$, after a dilation with a scale factor of 2, are $N'(4, 4)$, $O'(8, 4)$, and $P'(6, 10)$. The graph is shown in Step 2.**

EXAMPLE 2

Find the coordinates of the vertices of triangle FGH after a dilation with a scale factor of $\frac{1}{3}$. Then graph the dilation, triangle $F'G'H'$.

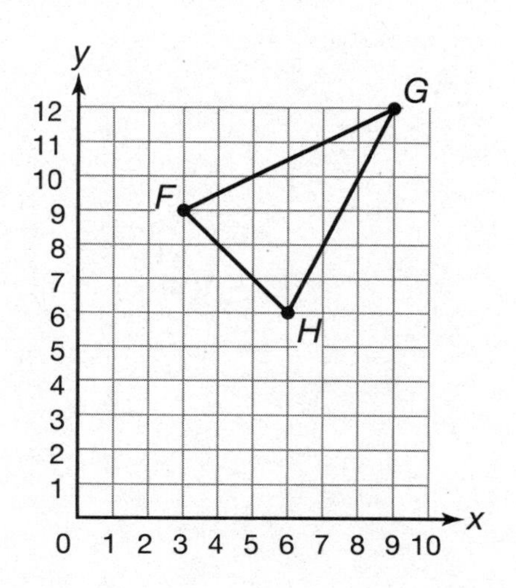

STRATEGY **Use the scale factor to identify the coordinates of the vertices of the image.**

STEP 1 Use the scale factor.

The scale factor is $\frac{1}{3}$, so multiply each coordinate of each vertex by $\frac{1}{3}$.

$F(3, 9) \rightarrow F'(3 \times \frac{1}{3}, 9 \times \frac{1}{3}) \rightarrow F'(1, 3)$

$G(9, 12) \rightarrow G'(9 \times \frac{1}{3}, 12 \times \frac{1}{3}) \rightarrow G'(3, 4)$

$H(6, 6) \rightarrow H'(6 \times \frac{1}{3}, 6 \times \frac{1}{3}) \rightarrow H'(2, 2)$

STEP 2 Graph $F'G'H'$.

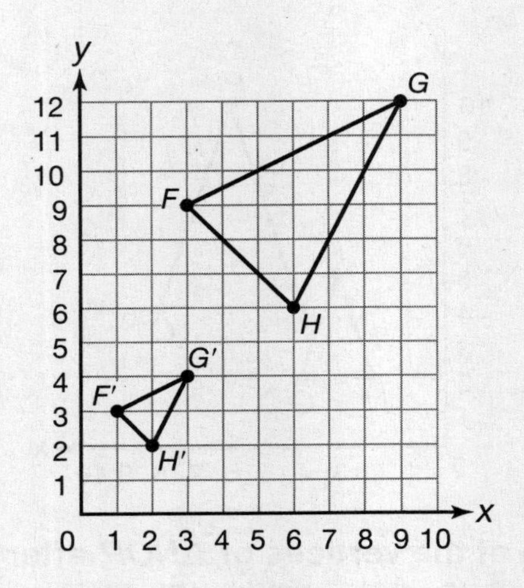

SOLUTION **The coordinates of the vertices of triangle *FGH* after a dilation with a scale factor of $\frac{1}{3}$ are $F'(1, 3)$, $G'(3, 4)$, $H'(2, 2)$. The graph is shown in Step 2.**

EXAMPLE 3

Rectangle *JKLM* and its image after a dilation are shown. What is the scale factor of the dilation?

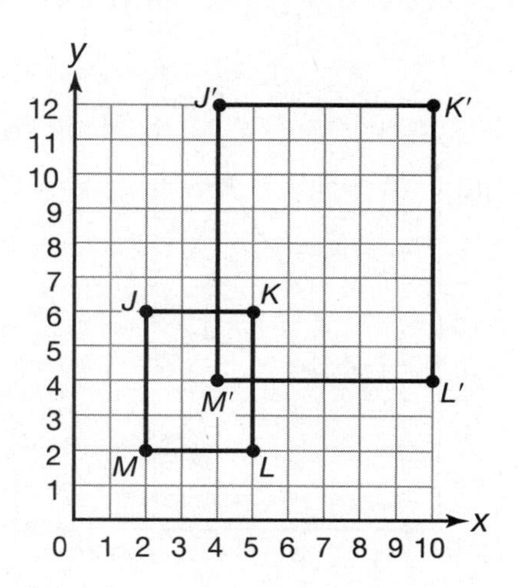

STRATEGY **Compare coordinates.**

The coordinates of *J* are (2, 6), and the coordinates of *J′* are (4, 12).

The coordinates of *J′* are 2 times the coordinates of *J*.

The coordinates of *K* are (5, 6), and the coordinates of *K′* are (10, 12).

The coordinates of *K′* are 2 times the coordinates of *K*.

The coordinates of *L* are (5, 2), and the coordinates of *L′* are (10, 4).

The coordinates of *L′* are 2 times the coordinates of *L*.

The coordinates of *M* are (2, 2), and the coordinates of *M′* are (4, 4).

The coordinates of *M′* are 2 times the coordinates of *M*.

SOLUTION **The coordinates of the image's vertices are all 2 times the coordinates of the original vertices, so the scale factor of the dilation is 2.**

Note: The scale factor of 2 in Example 3 also applies to the ratio of the lengths of the corresponding sides of *JKLM* and *J′K′L′M′*. For example, side *JM* in the original figure is 4 units long. Side *J′M′* in the image is 8 units long, or 2 times the length of *JM*.

COACHED EXAMPLE

Find the coordinates of the vertices of triangle *XYZ* after a dilation of $\frac{1}{2}$. Then graph the dilation.

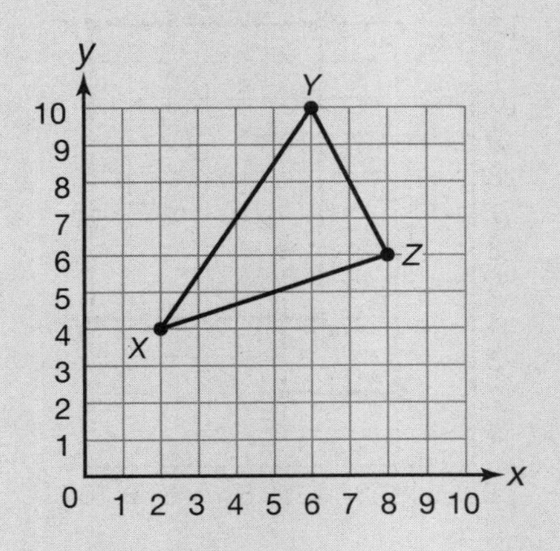

THINKING IT THROUGH

The coordinates of vertex *X* are __________.

The coordinates of vertex *Y* are __________.

The coordinates of vertex *Z* are __________.

The scale factor is ______, so multiply both coordinates of each vertex by ______.

Multiply to find the coordinates of each vertex of the image triangle $X'Y'Z'$.

X'(______ × ______, ______ × ______) → (______, ______)

Y'(______ × ______, ______ × ______) → (______, ______)

Z'(______ × ______, ______ × ______) → (______, ______)

The coordinates of the vertices of triangle *XYZ* after a dilation of $\frac{1}{2}$ are

____

Graph triangle $X'Y'Z'$ on the coordinate grid with triangle *XYZ*.

The graph of $X'Y'Z'$ is shown above.

Lesson Practice

Choose the correct answer.

1. Which transformation will **not** always produce an image that is congruent to the original figure?

   A. translation
   B. rotation
   C. reflection
   D. dilation

2. A dilation of triangle ABC has a scale factor of 4. If the coordinates of the vertices are $A(4, 8)$, $B(12, 8)$, and $C(16, 20)$, what are the coordinates of the vertices of the image after the dilation?

   A. A' (16, 32), B' (48, 32), C' (64, 80)
   B. A' (8, 12), B' (16, 12), C' (20, 24)
   C. A' (1, 2), B' (3, 2), C' (4, 5)
   D. A' (0, 4), B' (8, 4), C' (12, 16)

3. The area of rectangle $PQRS$ is 48 square inches. Under which kind of transformation could the area of its image, rectangle $P'Q'R'S'$, be less than 48 square inches?

   A. dilation
   B. reflection
   C. rotation
   D. translation

4. Rectangle $W'X'Z'Y'$ is the image of rectangle $WXZY$ after a dilation. What is the scale factor of the dilation?

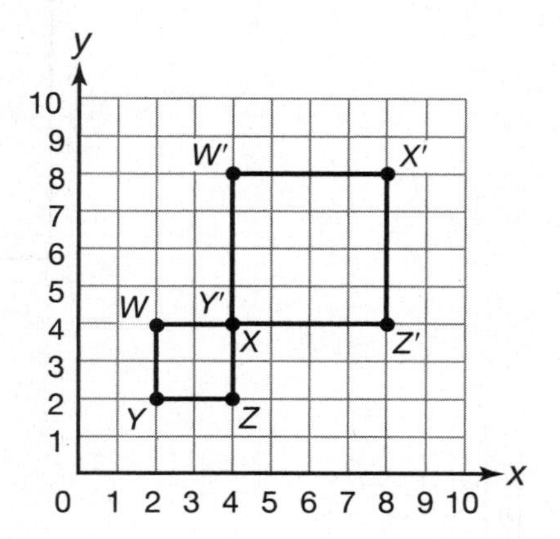

   A. 0.5
   B. 1.5
   C. 2
   D. 2.5

5. Which of the following shows the image of triangle RST after a dilation with a scale factor of 2?

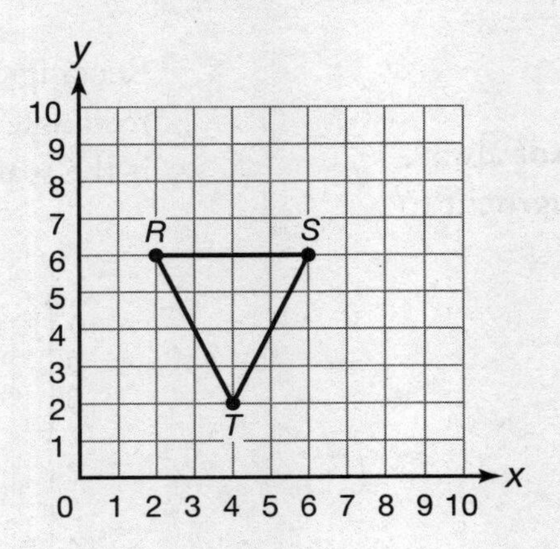

A.

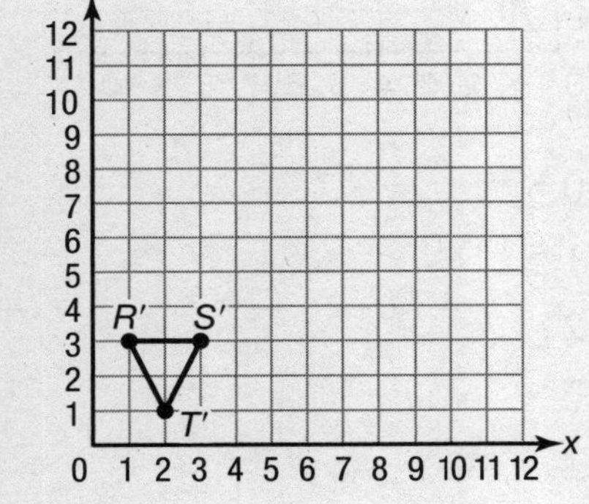

B.

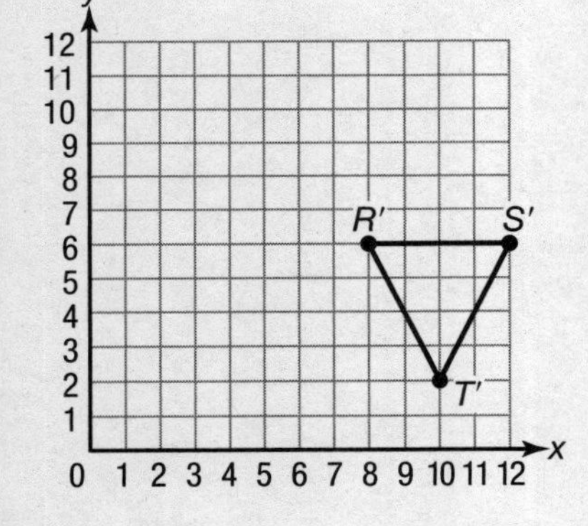

C.

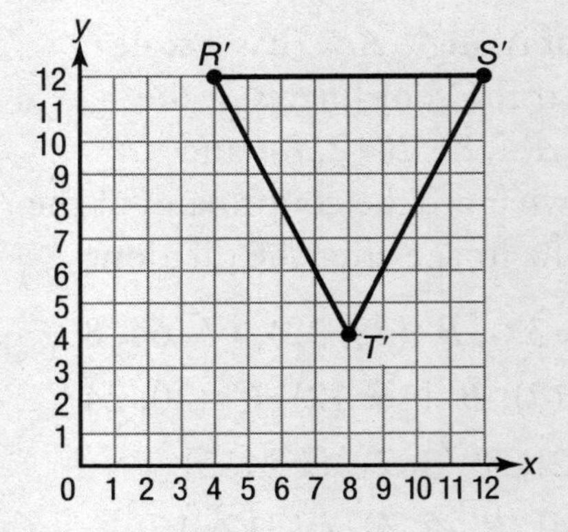

D.

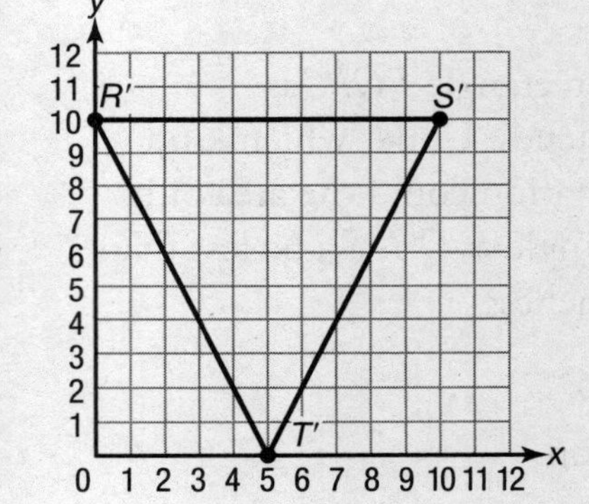

EXTENDED-RESPONSE QUESTION

6. Triangle *ABC* has vertices $A(1, 6)$, $B(5, 4)$, and $C(3, 1)$.

Part A What are the coordinates of the vertices of the image of triangle *ABC* after a dilation with a scale factor of 3?

Answer ____________

Part B On the coordinate grid below, graph triangle *ABC* and its image after a dilation with a scale factor of 3.

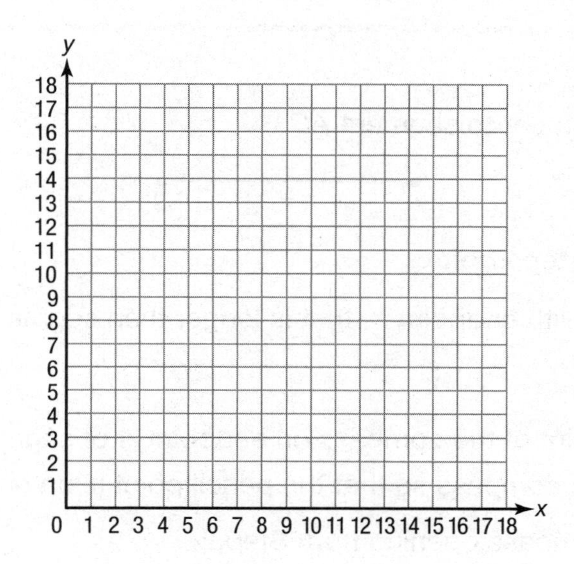

Lesson 22 Construct Geometric Figures

8.G.0

Getting the Idea

A **construction** is a precise way of drawing that allows only two tools: a straightedge and a compass.

EXAMPLE 1

Construct a segment congruent to segment *AB*.

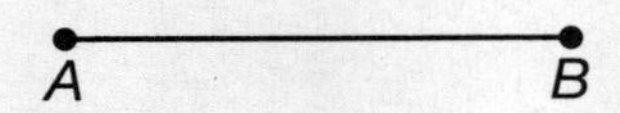

STRATEGY **Follow the steps below.**

STEP 1 Draw a ray, with endpoint *X*, that is longer than segment *AB*.

STEP 2 Place the point of the compass on endpoint *A* of segment *AB*.

Open the compass so that the pencil point is on endpoint *B*.

STEP 3 Keep the compass opening from Step 2.

Place the point of the compass on endpoint *X* of the ray. Then draw an arc that intersects the ray. Label the point of intersection *Y*.

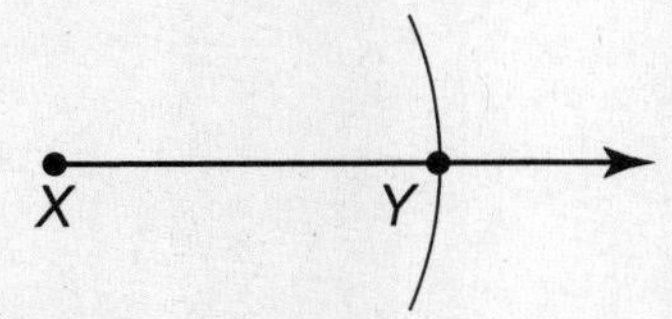

SOLUTION **Step 3 shows the construction of segment *XY* congruent to segment *AB*.**

EXAMPLE 2

Construct an angle congruent to $\angle A$.

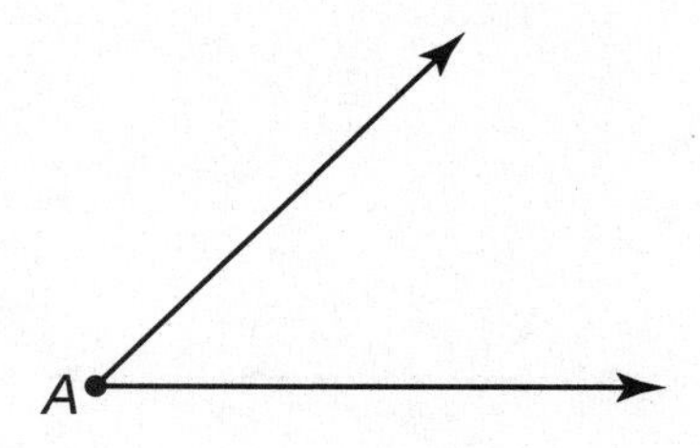

STRATEGY **Follow the steps below.**

STEP 1 Place the point of the compass on vertex A.

Draw an arc that intersects both rays of the angle. Label the points of intersection B and C.

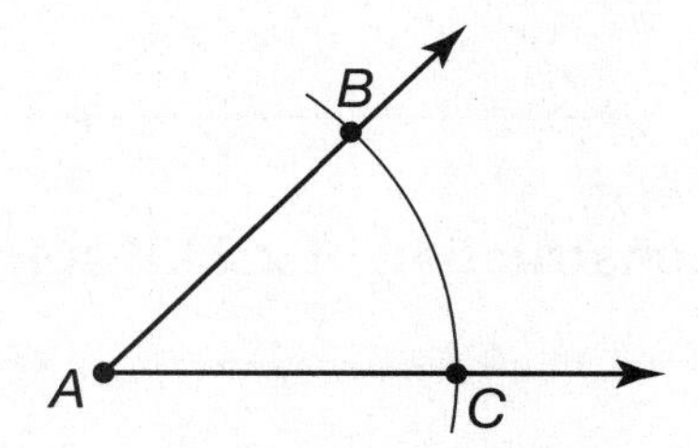

STEP 2 Draw a ray with endpoint X. Use the compass opening from Step 1.

Place the compass point on X and draw an arc that intersects the ray. Label the point of intersection Y.

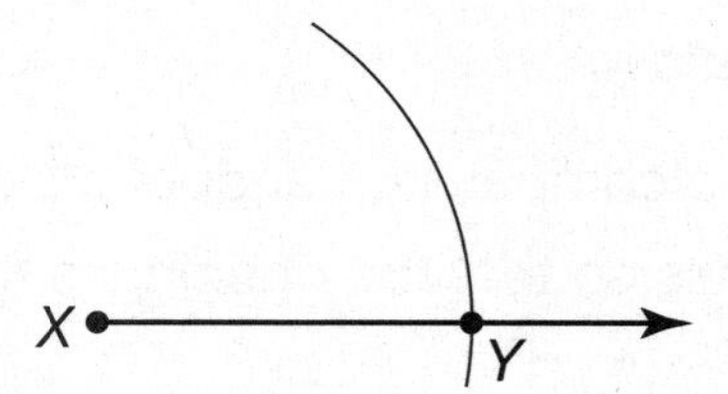

STEP 3 Place the compass point on point C on $\angle A$.

Open the compass until the pencil point is on point B. Draw an arc through point B.

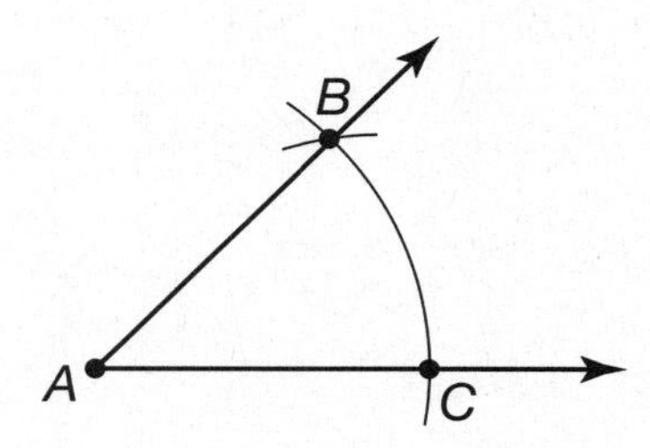

STEP 4 Keep the compass opening from Step 3.

Place the compass point on point Y and draw an arc that intersects the arc drawn in Step 2. Label the point of intersection Z.

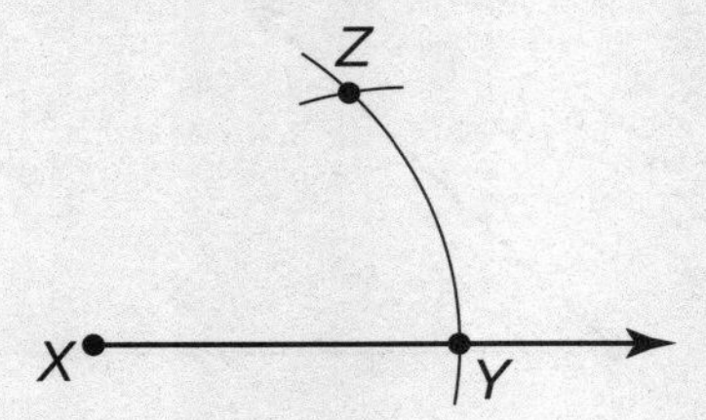

STEP 5 Draw a ray from point X through point Z.

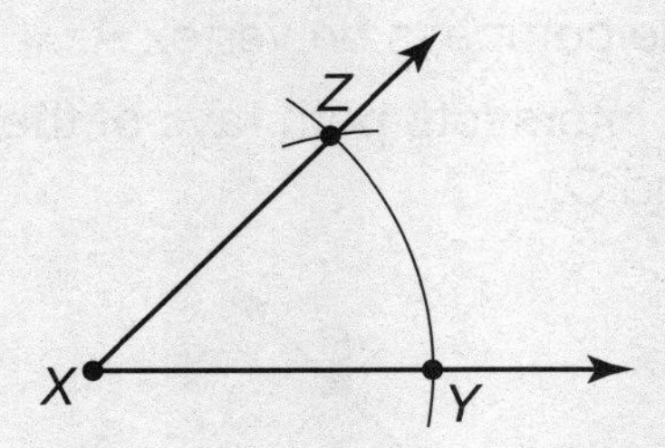

SOLUTION Step 5 shows the construction of $\angle ZXY$ congruent to $\angle A$.

EXAMPLE 3

Construct the perpendicular bisector of segment DE.

STRATEGY Follow the steps below.

STEP 1 Place the point of the compass on endpoint D.

Open the compass so that the opening is more than half the distance from D to E. Draw an arc.

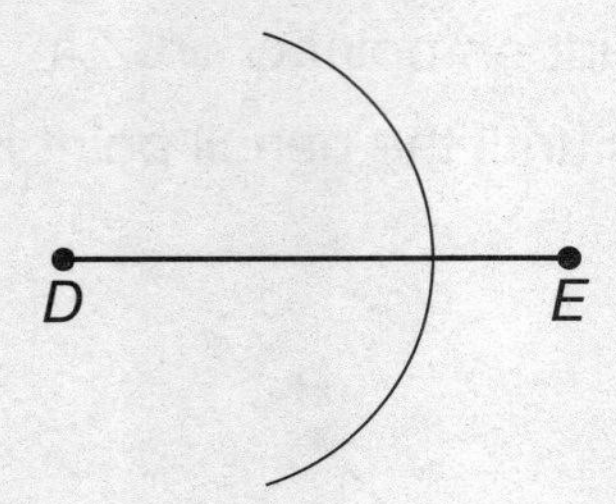

STEP 2 Keep the compass opening from Step 1.

Place the compass point on endpoint E and draw another arc that intersects the arc drawn in Step 1 twice. Label the points of intersection F and G.

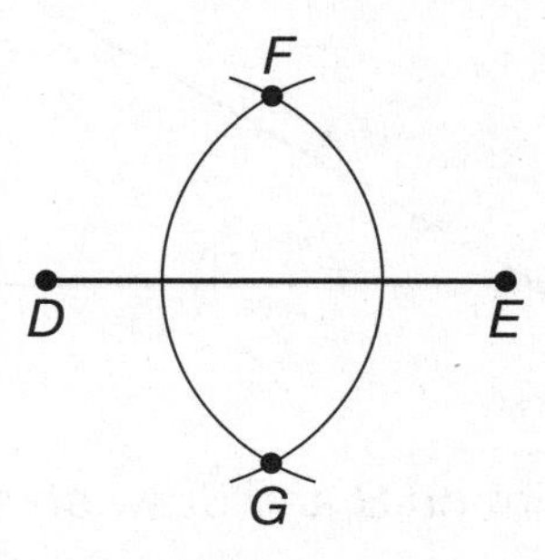

STEP 3 Draw a line through points F and G.

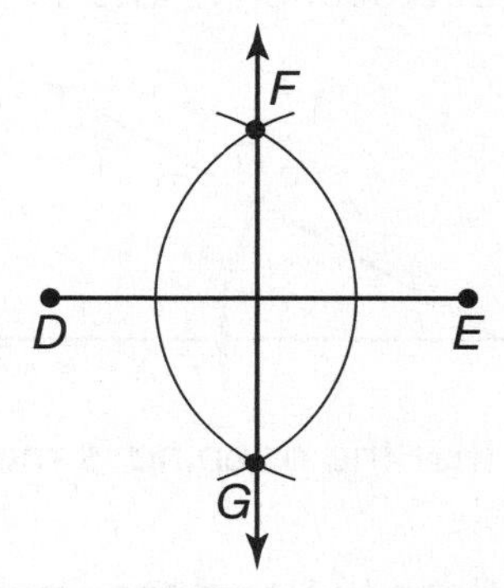

SOLUTION **Step 3 shows the construction of line *FG*, the perpendicular bisector of segment *DE*.**

EXAMPLE 4

Construct the angle bisector of $\angle S$.

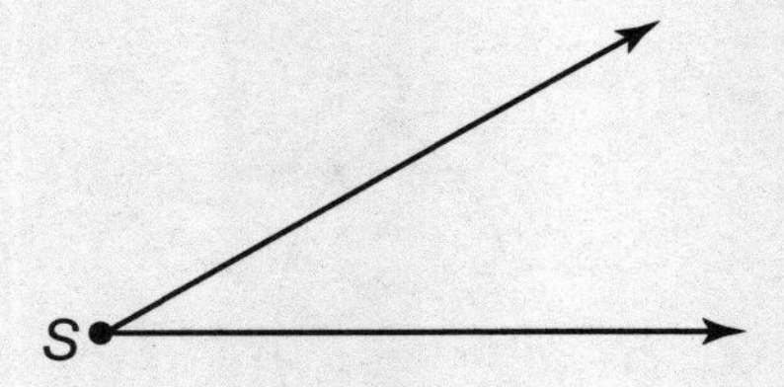

STRATEGY **Follow the steps below.**

STEP 1 Place the compass point on S and draw an arc that intersects the rays of the angle.

Label the points of intersection R and T.

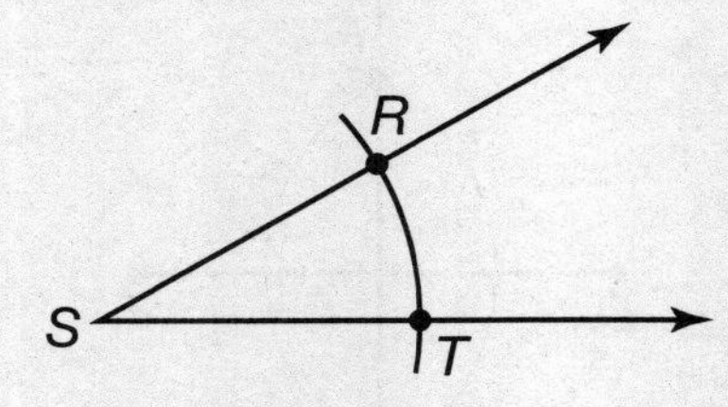

STEP 2 Open the compass so that the opening is more than half the distance from point R to point T.

Using this setting, place the compass point on T and draw an arc in the interior of the angle. Using the same setting, place the compass point on R and draw an arc that intersects the interior arc. Label the point of intersection V.

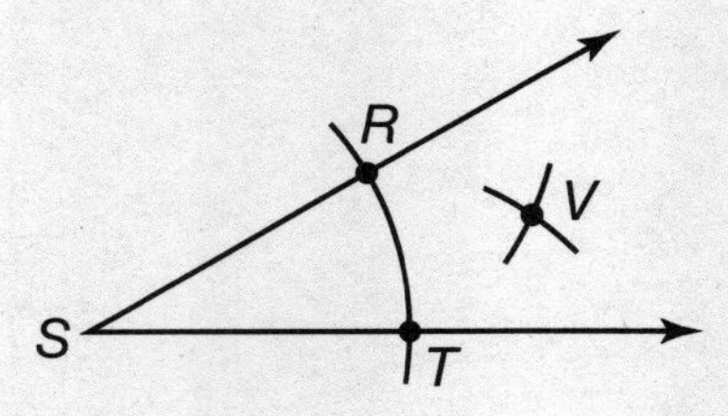

STEP 3 Draw ray SV.

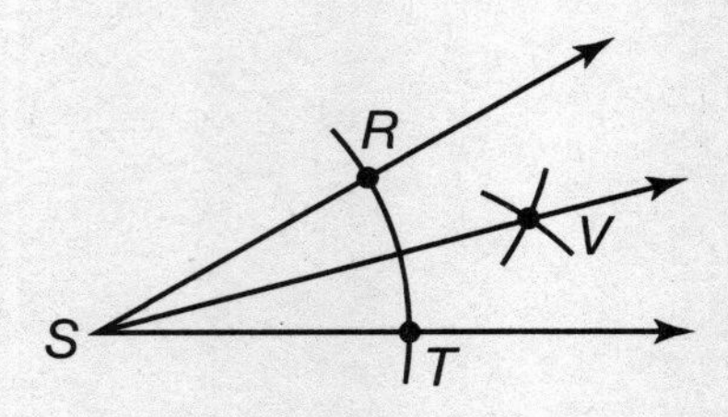

SOLUTION **Step 3 shows the construction of ray SV, the bisector of $\angle S$.**

COACHED EXAMPLE

Construct the perpendicular bisector of segment *PQ* below.

THINKING IT THROUGH

Place the compass point on endpoint *P* and open the compass so that the setting is greater than ____________ the distance from ____________ to ____________.

Draw an arc.

Using the same compass opening, place the compass point on endpoint ____________ and draw an arc that intersects the first arc twice.

Draw a ____________ through the intersection points of the arcs.

The line is the perpendicular bisector of segment *PQ*.

P *Q*

The construction of the perpendicular bisector of *PQ* is shown above.

Lesson Practice

Choose the correct answer.

1. Which kind of construction is shown below?

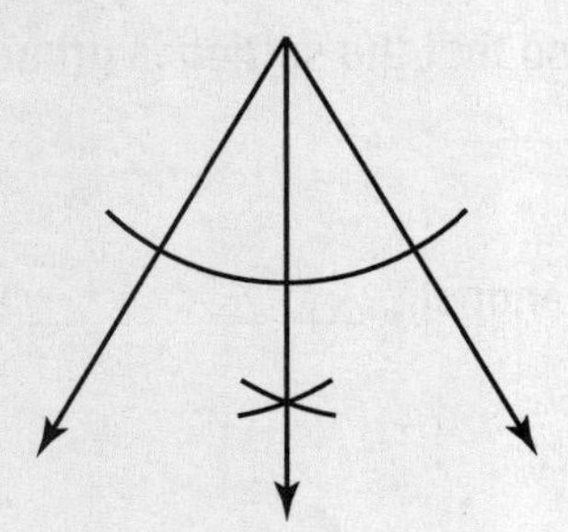

A. a segment congruent to a given segment

B. an angle congruent to a given angle

C. the bisector of an angle

D. the perpendicular bisector of a segment

2. Which kind of construction is shown below?

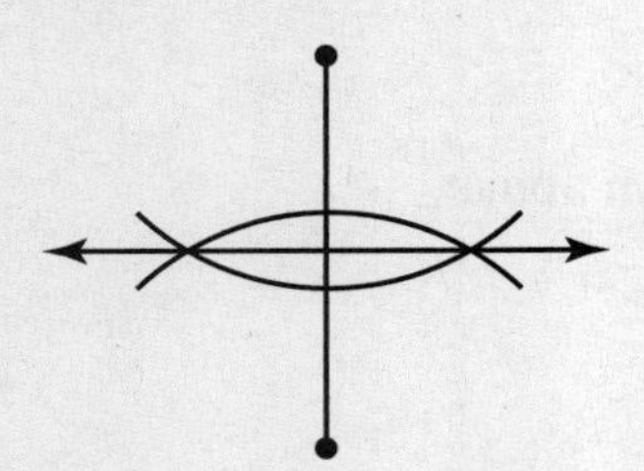

A. a segment congruent to a given segment

B. an angle congruent to a given angle

C. the bisector of an angle

D. the perpendicular bisector of a segment

3. Neil wants to construct the bisector of $\angle JKL$. What should he do first?

A. Place the compass point on *J* and draw an arc.

B. Place the compass point on *K* and draw an arc.

C. Place the compass point on *L* and draw an arc.

D. Use a ruler to measure the distance from *J* to *L*.

4. Wendy wants to construct a segment congruent to a given segment *MN*. What should she do first?

A. Draw a ray on which to construct the segment.

B. Find the midpoint of segment *MN*.

C. Draw an angle with point *M* as the vertex.

D. Use a ruler to measure the length of segment *MN*.

5. Leah wants to construct the perpendicular bisector of segment WX. What should she do first?

A. Set the compass opening so that it is greater than half the distance from W to X and draw an arc.

B. Set the compass opening so that it is equal to half the distance from W to X and draw an arc.

C. Set the compass opening so that it is less than half the distance from W to X and draw an arc.

D. Construct a segment congruent to segment WX.

6. Kevin wants to construct an angle congruent to $\angle NOM$. What should he do first?

A. Place the compass point on point M and draw an arc.

B. Place the compass point on point N and draw an arc.

C. Place the compass point on point O and draw an arc.

D. Measure the distance from point N to point M.

EXTENDED-RESPONSE QUESTION

7. Use a compass and a straightedge.

Part A Construct the bisector of $\angle D$. Show the arcs you make.

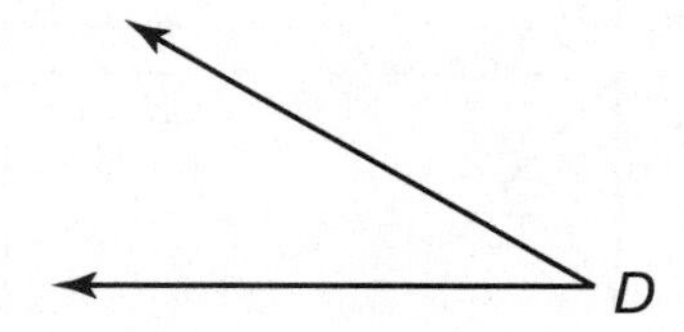

Part B Explain how you constructed the angle bisector.

___

___

___

Lesson

23 Slope

8.G.13

Getting the Idea

You can think of the **slope** of a line as the steepness of the line, or the ratio $\frac{\text{rise}}{\text{run}}$.

To find the slope, label any two points on a line as (x_1, y_1) and (x_2, y_2). You can then find the slope, m, using this ratio:

$$\text{slope} = m = \frac{\text{change in } y}{\text{change in } x} = \frac{y_2 - y_1}{x_2 - x_1}$$

Look at the lines graphed below.

- A line that slants up from left to right has a positive slope.
- A line that slants down from left to right has a negative slope.
- A horizontal line has a slope of 0, because it does not rise.
- A vertical line has an undefined slope because the change in its x-values is 0, and it is impossible to divide by 0.

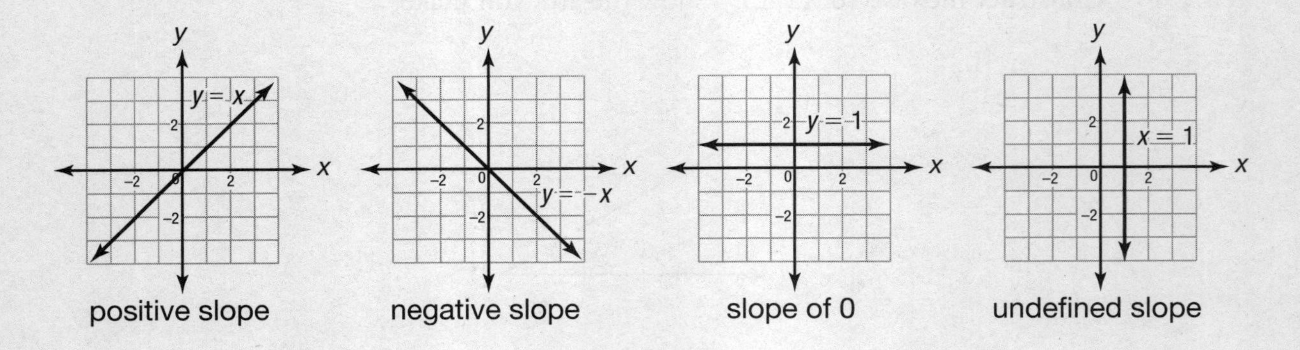

positive slope negative slope slope of 0 undefined slope

EXAMPLE 1

What is the slope of this line?

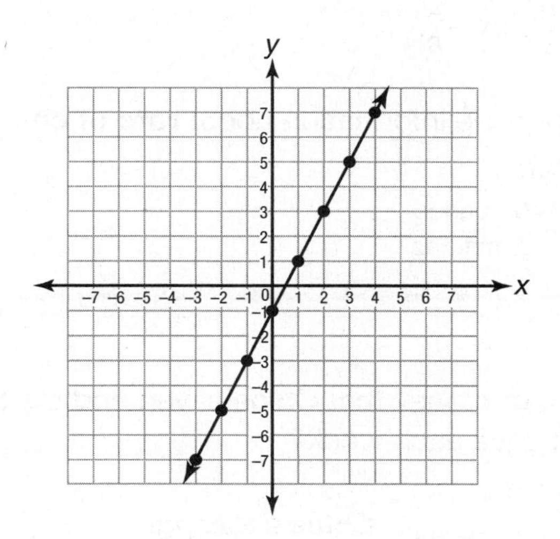

STRATEGY **Find the changes in *x* and *y*.**

STEP 1 Use the coordinates of any two points on the line.

Let $(2, 3) = (x_1, y_1)$.

Let $(4, 7) = (x_2, y_2)$.

STEP 2 Find the change in *y*.

$y_2 - y_1 = 7 - 3 = 4$

STEP 3 Find the change in *x*.

$x_2 - x_1 = 4 - 2 = 2$

STEP 4 Form a ratio.

$\frac{\text{rise}}{\text{run}} = \frac{4}{2}$ or $\frac{2}{1}$

SOLUTION **The slope of the line is 2.**

You can find the slope if you are given two points on a line.

EXAMPLE 2

What is the slope of the line that passes through (−3, 6) and (5, 1)?

STRATEGY **Find $\frac{\textbf{change in } y}{\textbf{change in } x}$.**

STEP 1 Find the change in *y*.

$1 - 6 = -5$

STEP 2 Find the change in x.

$$5 - (-3) = 8$$

SOLUTION **The slope is $\frac{-5}{8}$, or $-\frac{5}{8}$.**

The slope of a line can also provide information about **rate of change**. For example, a slope of $\frac{50}{3}$ could represent these rates:

$\frac{50 \text{ miles}}{3 \text{ gallons}}$ $\frac{\$50}{3 \text{ hours}}$ $\frac{50 \text{ pages}}{3 \text{ minutes}}$

EXAMPLE 3

This graph shows the amount of money that Chloe saved and the time it took her to save the money. How much did Chloe save each week?

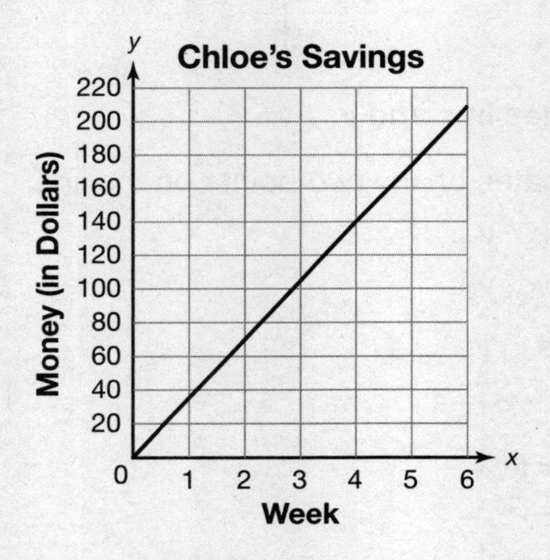

STRATEGY **Find the slope of the line to find the rate.**

STEP 1 Use the coordinates of any two points on the line.

Let $(0, 0) = (x_1, y_1)$.

Let $(4, 140) = (x_2, y_2)$.

STEP 2 Use the slope formula. Substitute the numbers.

$$m = \frac{y_2 - y_1}{x_2 - x_1}$$

$$m = \frac{140 - 0}{4 - 0}$$

$$m = \frac{140}{4} = 35$$

STEP 3 Interpret the slope.

$$\text{slope} = \frac{35}{1} = \frac{\text{money in dollars}}{\text{number of weeks}}$$

SOLUTION **Chloe saved \$35 per week.**

COACHED EXAMPLE

The graph below shows the gallons of gas a hybrid bus had used after traveling certain distances. The gas mileage of this bus is measured in miles per gallon. What is the bus's gas mileage?

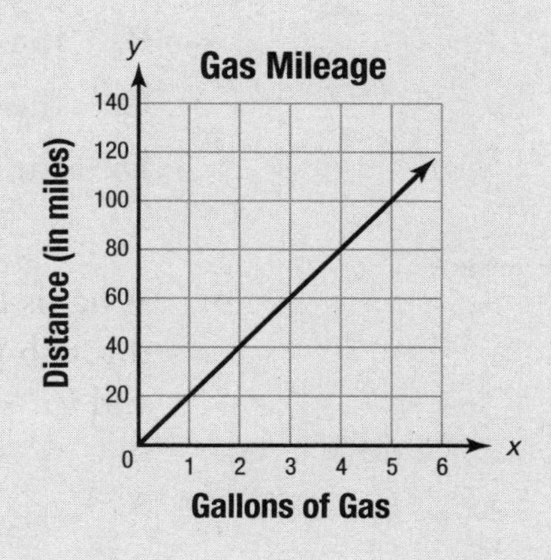

THINKING IT THROUGH

The gas mileage is measured in miles per gallon. So, gas mileage is a __________.

The __________ of the graph represents the bus's gas mileage.

Let's choose two points on the graph. Let (x_1, y_1) be (1, ______), and (x_2, y_2) be (3, ______).

Find the slope. $m = \frac{y_2 - y_1}{x_2 - x_1} =$ __________

The bus's gas mileage is __________ miles per gallon.

Lesson Practice

Choose the correct answer.

1. What is the slope of this line?

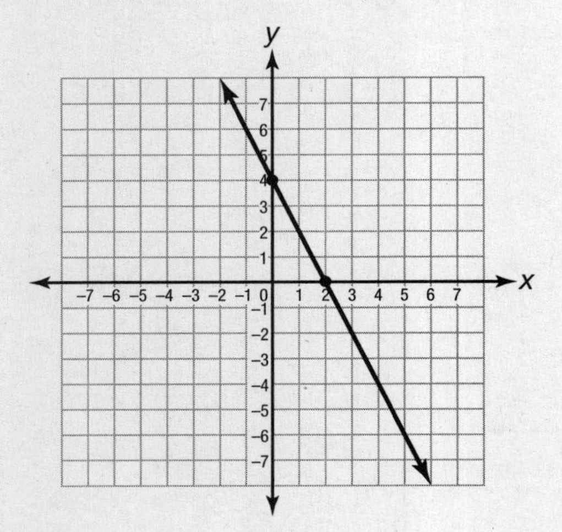

A. -2

B. $-\frac{1}{2}$

C. $\frac{1}{2}$

D. 2

2. What is the slope of this line?

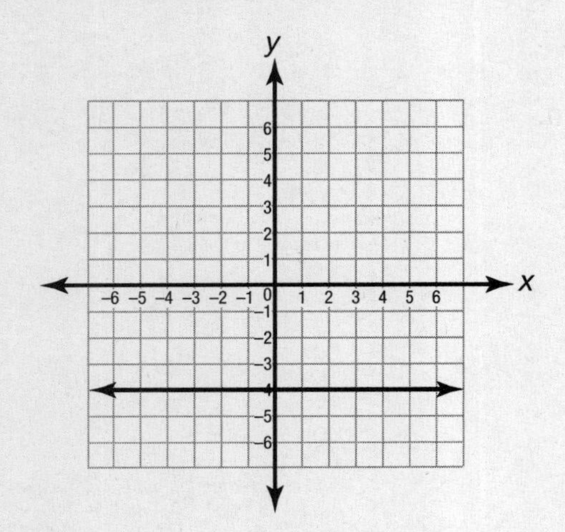

A. -4

B. 0

C. 4

D. The slope is undefined.

3. The graph of a line contains points at (3, 1) and (2, 5). Which of the following must be true about the graph of this line?

A. The slope is negative.

B. The slope is 0.

C. The slope is positive.

D. The slope is undefined.

4. What is the slope of a line that passes through points with coordinates $(2, -5)$ and $(6, -2)$?

A. $-\frac{4}{3}$

B. $-\frac{3}{4}$

C. $\frac{3}{4}$

D. $\frac{4}{3}$

5. What is the slope of a line that passes through points with coordinates $(-4, 2)$ and $(6, -4)$?

A. $-\frac{5}{3}$

B. $-\frac{3}{5}$

C. $\frac{3}{5}$

D. $\frac{5}{3}$

6. What is the slope of a line that passes through points with coordinates $(-3, 4)$ and $(-3, -2)$?

A. -1

B. 0

C. 1

D. The slope is undefined.

7. This graph shows the number of hours Ginny cycled and the distance she traveled.

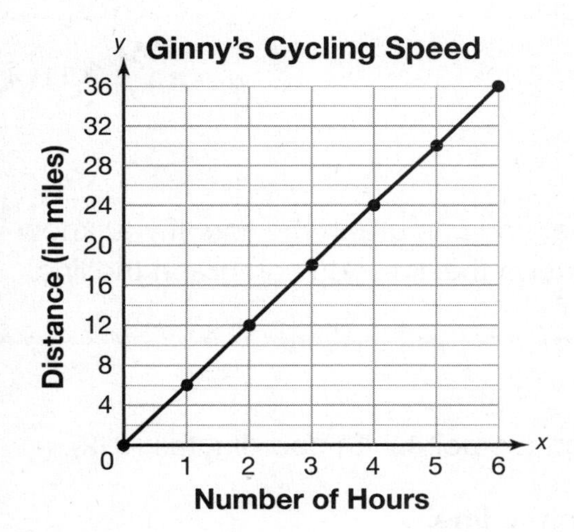

What is Ginny's average cycling speed?

A. 6 miles per hour

B. 8 miles per hour

C. 12 miles per hour

D. 16 miles per hour

8. Manuel drives a pedicab in New York City. This graph shows the rates he charges.

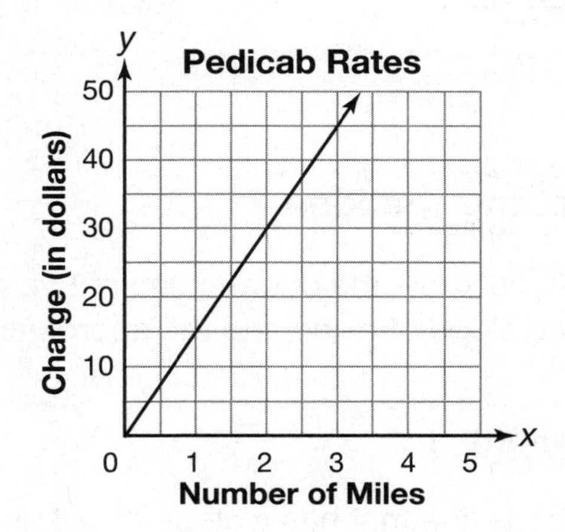

What is his rate, in dollars per mile?

Answer ____________

Lesson

24 Determine the y-Intercept

8.G.14, 8.G.15

Getting the Idea

If you know the coordinates of two points on a line, you can graph the line. If you know the slope of a line and the coordinates of one point on the line, you can graph the line.

EXAMPLE 1

Graph the line that has a slope of -2 and goes through the point with coordinates (4, 2).

STRATEGY **Use the slope to find another point on the line.**

STEP 1 What does a slope of -2 mean?

Think: $-2 = -\frac{2}{1}$

The value of y decreases by 2 and the value of x increases by 1.

STEP 2 Find another point on the line.

Start with (4, 2).

Decrease the y-value by 2: $2 - 2 = 0$

Increase the x-value by 1: $4 + 1 = 5$

Another point on the line is (5, 0).

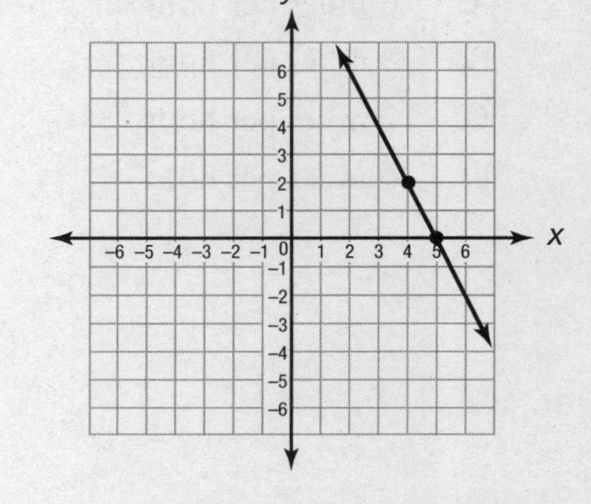

STEP 3 Plot the points and draw the line.

SOLUTION **The graph of the line that has a slope of -2 and goes through the point with coordinates (4, 2) is shown in Step 3.**

The ***x*-intercept** of a line is the x-coordinate of the point where a line intersects the x-axis. The value of the y-coordinate of this point is always 0.

The ***y*-intercept** of a line is the y-coordinate of the point where a line intersects the y-axis. The value of the x-coordinate of this point is always 0.

EXAMPLE 2

Write the ordered pairs for the *x*-intercept and for the *y*-intercept of the line in the graph below.

STRATEGY **Identify the points where the line intersects the *x*-axis and the *y*-axis.**

STEP 1 Identify the *x*-intercept.

The *x*-coordinate of the *x*-intercept is −4.
The *y*-coordinate of this point must be 0.

The ordered pair for this point is (−4, 0).

STEP 2 Identify the *y*-intercept.

The *y*-coordinate of the *y*-intercept is −3. The *x*-coordinate of this point must be 0.

The ordered pair for this point is (0, −3).

SOLUTION **The ordered pair for the *x*-intercept is (−4, 0); the ordered pair for the *y*-intercept is (0, −3).**

EXAMPLE 3

Identify the slope, the *x*-intercept, and the *y*-intercept of this graph.

STRATEGY **Find the slope and intercepts of the graph.**

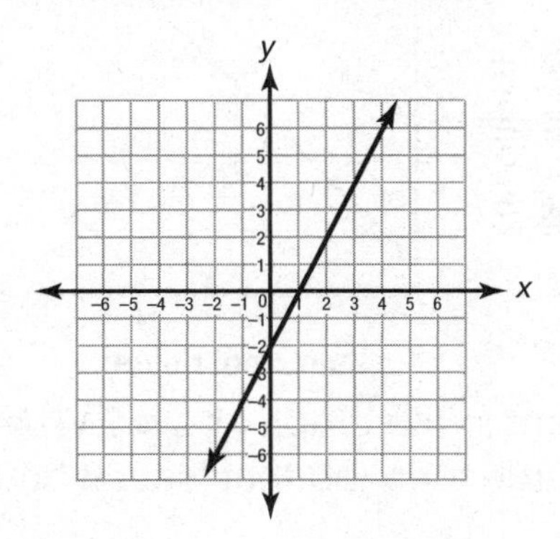

STEP 1 Find the slope.

Choose two points whose coordinates are numbers that are easy to work with.

For example, let (x_1, y_1) be (2, 2), and let (x_2, y_2) be (3, 4).

Find the slope, *m*:

$$m = \frac{y_2 - y_1}{x_2 - x_1} = \frac{4 - 2}{3 - 2} = \frac{2}{1} = 2$$

STEP 2 Find the x-intercept.

The line crosses the x-axis at (1, 0), so the x-intercept is 1.

STEP 3 Find the y-intercept.

The line crosses the y-axis at (0, −2), so the y-intercept is −2.

STEP 4 Show the x- and y-intercepts on the graph.

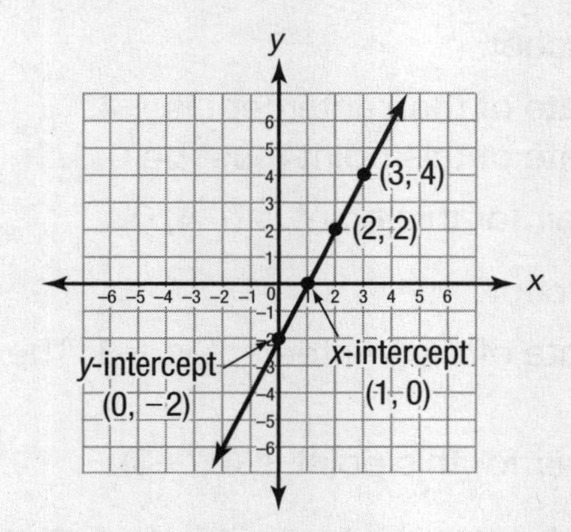

SOLUTION The slope is 2, the x-intercept is 1, and the y-intercept is −2.

EXAMPLE 4

A plumber charges a fee for each house call plus an hourly rate, as shown by the graph.

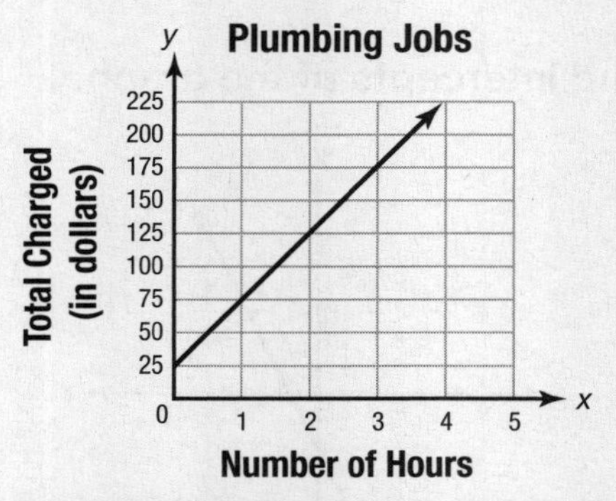

What does the y-intercept represent in the graph above? What does the slope represent? Use that information to predict the charge for a job that lasts 8 hours.

STRATEGY Identify the slope and the y-intercept.

STEP 1 Find the y-intercept.

The graph touches the y-axis at (0, 25).

That means that if the plumber works 0 hours, the fee is $25.

The y-intercept shows that the fee for the house call alone is $25.

STEP 2 Find the slope (rate of change).

Let (x_1, y_1) be (0, 25), and (x_2, y_2) be (1, 75).

$m = \frac{y_2 - y_1}{x_2} - x_1 = \frac{75 - 25}{1 - 0} = \frac{50}{1} = 50$

The *y*-values are dollars and the *x*-values are hours, so the rate of change is $\frac{\$50}{1 \text{ hour}}$, or \$50 per hour. The slope represents the hourly charge.

STEP 3 Predict the charge for a job that lasts 8 hours.

The fee is \$25 + \$50 per hour.

The charge is 25 + 50(8) = 25 + 400 = 425.

SOLUTION **The *y*-intercept represents the fee for a house call, \$25. The slope represents the amount charged per hour, \$50 per hour. The charge for an 8-hour job would be \$425.**

COACHED EXAMPLE

The graph of an equation is shown below.

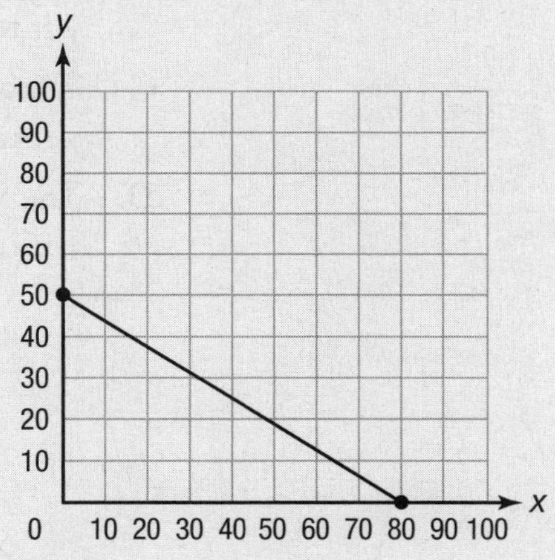

THINKING IT THROUGH

Find two points on the line: (__________, 50) and (80, __________)

Use those points to determine the slope.

$m = \frac{y_2 - y_1}{x_2 - x_1} = \frac{\_\_\_\_ - \_\_\_\_}{\_\_\_\_ - \_\_\_\_} = \_\_\_\_\_\_ = \_\_\_\_\_\_\_\_$

Now find the point where the graph intersects the *y*-axis.

The graph crosses the *y*-axis at (__________, __________), so the *y*-intercept is __________.

The slope is __________, and the *y*-intercept is __________.

Lesson Practice

Choose the correct answer.

1. What is the y-intercept of the line graphed below?

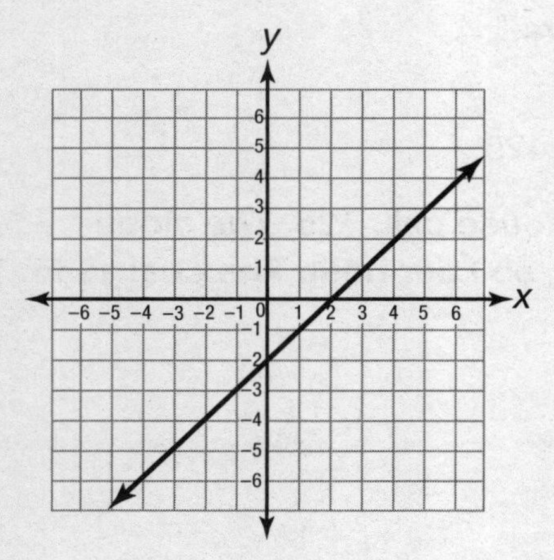

A. 2

B. 0

C. -2

D. -3

2. What are the x- and y-intercepts of this line?

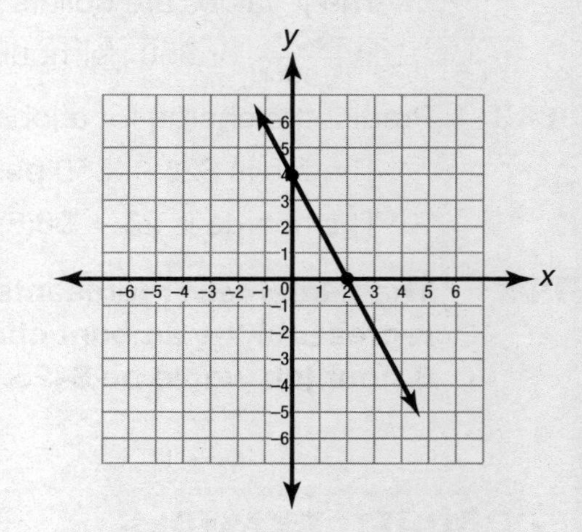

A. The x-intercept is 4, and the y-intercept is 2.

B. The x-intercept is 0, and the y-intercept is 0.

C. The x-intercept is 2, and the y-intercept is 4.

D. The x-intercept is -2, and the y-intercept is -4.

Use this information for questions 3 and 4.

A line has a slope of $\frac{1}{2}$ and passes through the point with coordinates $(-2, 1)$.

3. Which of the following is a table of values for the line?

A.

x	-6	-4	-2	0	2
y	-1	0	1	2	3

B.

x	-4	-3	-2	0	2
y	-3	-1	1	3	5

C.

x	-6	-4	-2	0	2
y	-3	-1	1	3	5

D.

x	-4	-3	-2	-1	0
y	-6	-3	1	3	6

4. Which of the following is the graph of the line?

A.

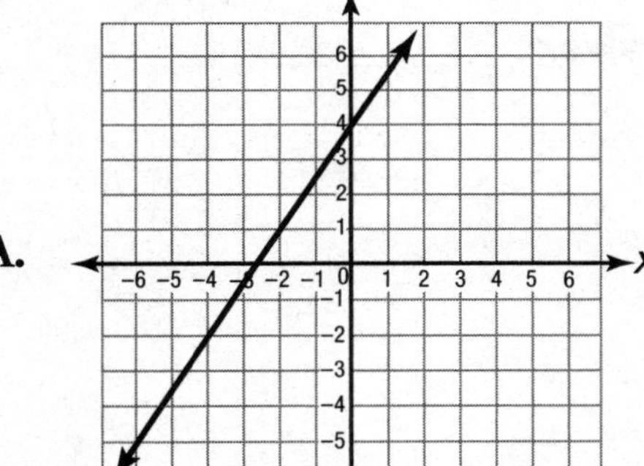

C.

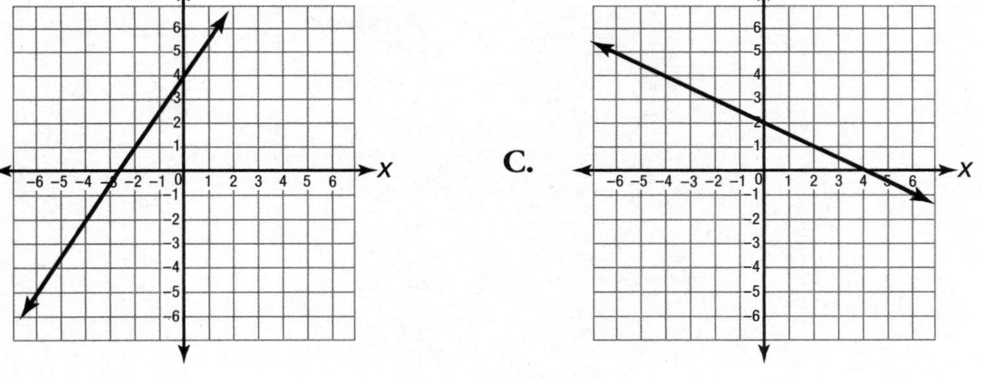

B.

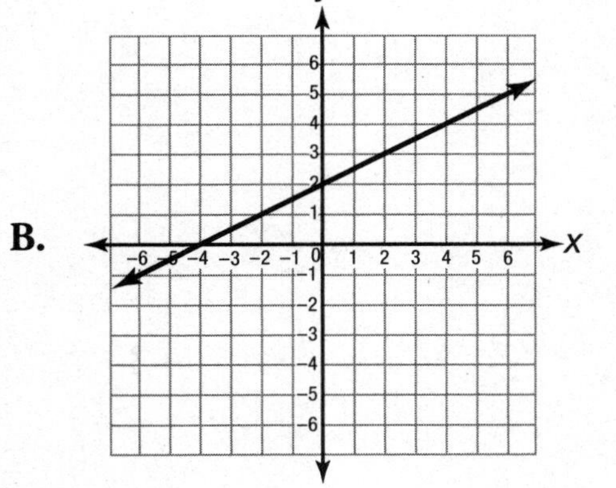

D.

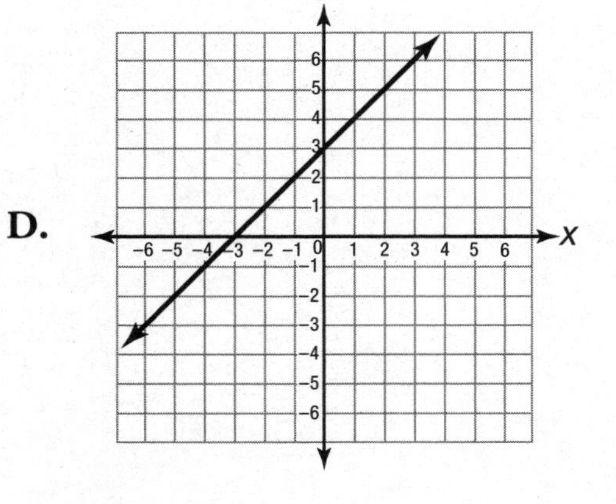

5. A cab charges a flat fee of \$2.50 and \$1 for each $\frac{1}{4}$ mile. The graph showing the cost of a ride is shown below.

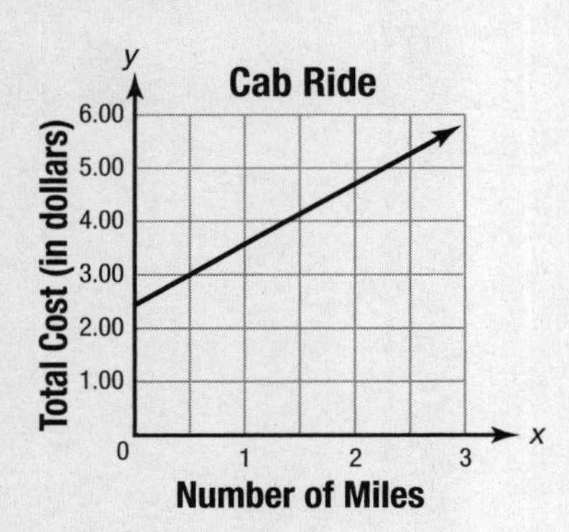

What does the y-intercept of the line represent?

A. the cost of a 1-mile ride

B. the cost of a 3-mile ride

C. the cost of hiring a cab

D. nothing

Use the table below for questions 6 and 7.

x	-4	-2	2	4
y	0	-1	-3	-4

6. Graph the line represented by the table of values.

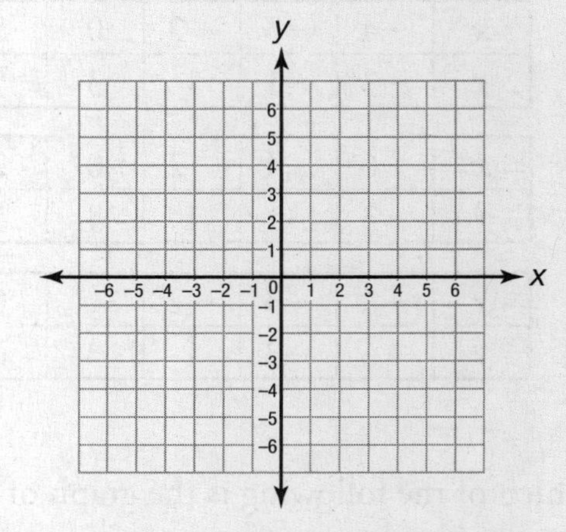

7. What is the y-intercept of the line?

Answer ________________

Lesson

25 Slope-Intercept Form

8.G.16, 8.G.17

Getting the Idea

The equation for a line can be given in **slope-intercept form**, $y = mx + b$. In that form, m represents the slope and b represents the y-intercept. If a linear equation is written in slope-intercept form, use the y-intercept and the slope to graph the equation.

EXAMPLE 1

Graph the equation $y = \frac{1}{2}x + 3$.

STRATEGY **Use the y-intercept and slope to graph the equation.**

STEP 1 Identify the slope and y-intercept.

In $y = \frac{1}{2}x + 3$, $m = \frac{1}{2}$ and $b = 3$.

The slope, m, is $\frac{1}{2}$.

The y-intercept, b, is 3, which is located at (0, 3).

STEP 2 Use the y-intercept and slope to find another point on the line.

Start with (0, 3).

Increase the x-value by 2: $0 + 2 = 2$

Increase the y-value by 1: $3 + 1 = 4$

Another point on the line is (2, 4).

STEP 3 Graph the equation.

SOLUTION **Step 3 shows the graph of $y = \frac{1}{2}x + 3$.**

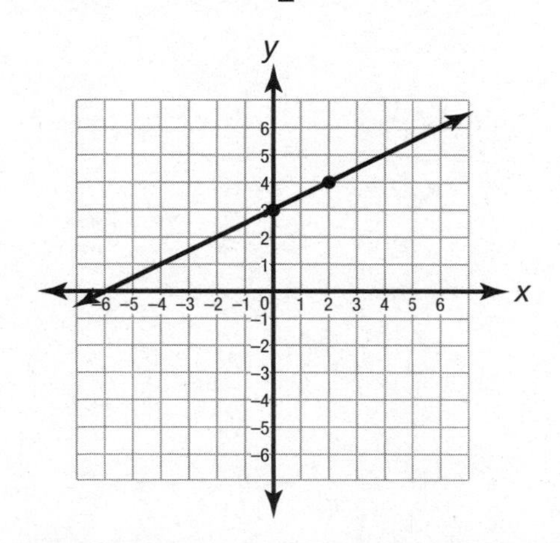

EXAMPLE 2

Graph the equation $y = -\frac{2}{3}x + 5$.

STRATEGY **Use the y-intercept and slope to graph the equation.**

STEP 1 Identify the slope and y-intercept.

In $y = -\frac{2}{3}x + 5$, $m = -\frac{2}{3}$ and $b = 5$.

The slope is $-\frac{2}{3}$ and the y-intercept is at (0, 5).

STEP 2 Use the slope to find a second point.

First, plot the point at the y-intercept, (0, 5).

Use the slope to find a second point.

$\text{slope} = \frac{\text{rise}}{\text{run}} = -\frac{2}{3}$

Start at (0, 5). Since the rise is negative, count 2 units down. Then count 3 units to the right, and plot a point there at (3, 3).

STEP 3 Draw a line through both points.

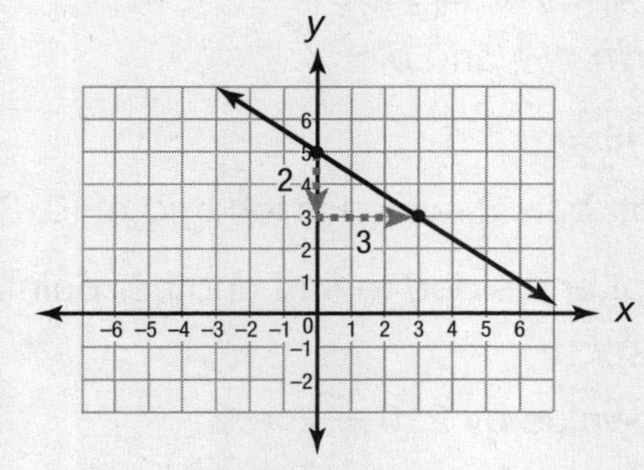

SOLUTION **The graph of $y = -\frac{2}{3}x + 5$ is shown in Step 3.**

EXAMPLE 3

Write an equation in slope-intercept form for this function.

x	−3	−2	−1	0	1
y	−2	0	2	4	6

STRATEGY **Identify the intercepts and then find the slope.**

STEP 1 Identify the x- and y-intercepts.

When $x = 0$, $y = 4$. The y-intercept is 4.

When $y = 0$, $x = -2$. The x-intercept is −2.

STEP 2 Use the intercepts to find the slope.

Use points (0, 4) and (−2, 0).

Change in y: $y_2 - y_1 = 0 - 4 = -4$

Change in x: $x_2 - x_1 = -2 - 0 = -2$

The slope is $\frac{-4}{-2}$, or 2.

STEP 3 Write the equation in slope-intercept form.

Slope-intercept form is $y = mx + b$.

The slope, m, is 2. The y-intercept, b, is 4.

SOLUTION **An equation for the function in slope-intercept form is $y = 2x + 4$.**

COACHED EXAMPLE

Write the equation of the line in slope-intercept form.

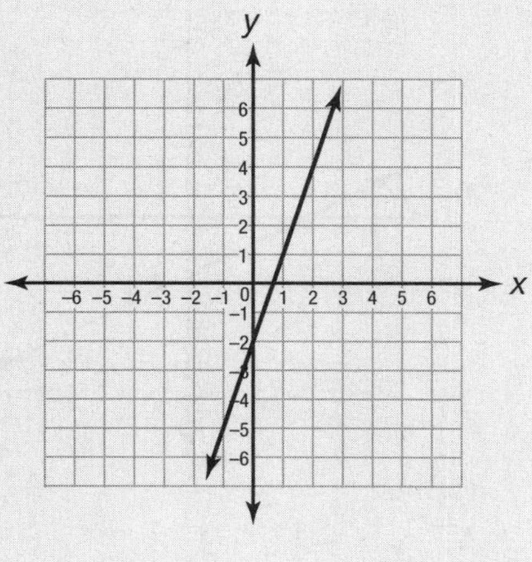

THINKING IT THROUGH

What is the y-intercept? __________

The slope is the ratio of $\frac{\text{change in }\_\_\_\_\_\_}{\text{change in }\_\_\_\_\_\_}$.

Pick any two points. ______________________________

Find the change in y. ______________________________

Find the change in x. ______________________________

The slope is __________.

To write the equation in slope-intercept form, use $y = mx + b$, where m is the __________ and b is the ______________________.

The equation of the line in slope-intercept form is ______________________.

Lesson Practice

Choose the correct answer.

1. What is the equation of this line in slope-intercept form?

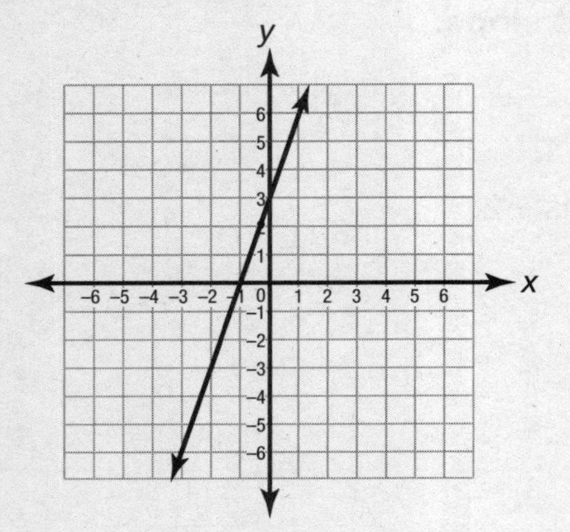

A. $y = x - 1$

B. $y = x + 3$

C. $y = 3x - 1$

D. $y = 3x + 3$

2. What is the equation of this line in slope-intercept form?

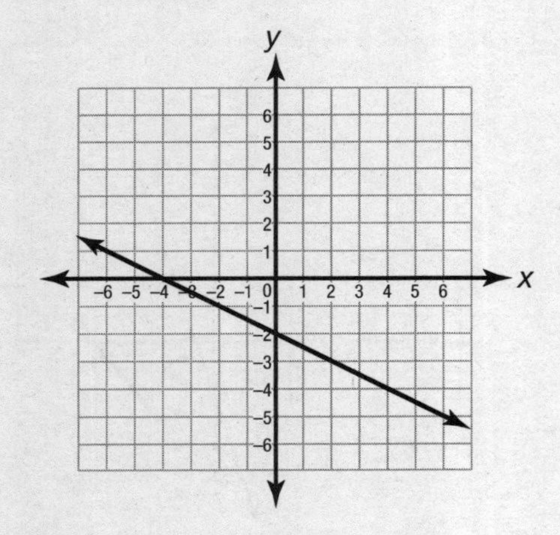

A. $y = -\frac{1}{2}x - 2$

B. $y = -\frac{1}{2}x - 4$

C. $y = -2x - 2$

D. $y = -2x - 4$

3. A line passes through points $(-6, 0)$ and $(0, 8)$ and has a slope of $\frac{4}{3}$. What is the equation of the line in slope-intercept form?

A. $y = \frac{4}{3}x - 6$

B. $y = \frac{4}{3}x + 6$

C. $y = \frac{4}{3}x - 8$

D. $y = \frac{4}{3}x + 8$

4. What is the equation in slope-intercept form of the function shown in this table?

x	−2	−1	0	1	2
y	0	3	6	9	12

A. $y = x - 2$

B. $y = x + 6$

C. $y = 3x - 2$

D. $y = 3x + 6$

5. What is the equation in slope-intercept form of the function shown in this table?

x	−2	0	2	4	6
y	3	2	1	0	−2

A. $y = -\frac{1}{2}x + 2$

B. $y = -\frac{1}{2}x + 4$

C. $y = -2x + 2$

D. $y = -2x + 4$

6. Which is the graph of the equation $y = \frac{2}{3}x - 2$?

A.

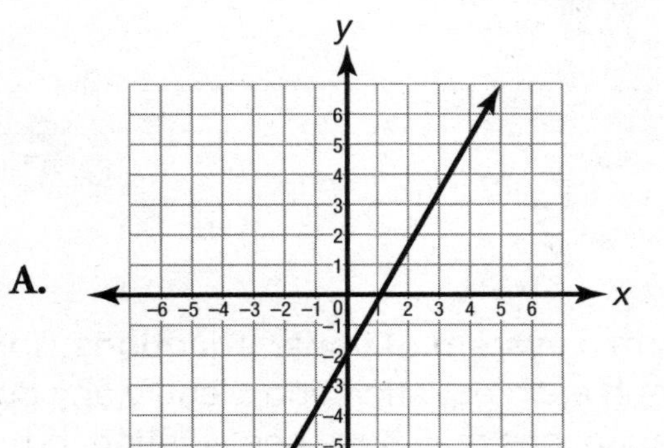

C.

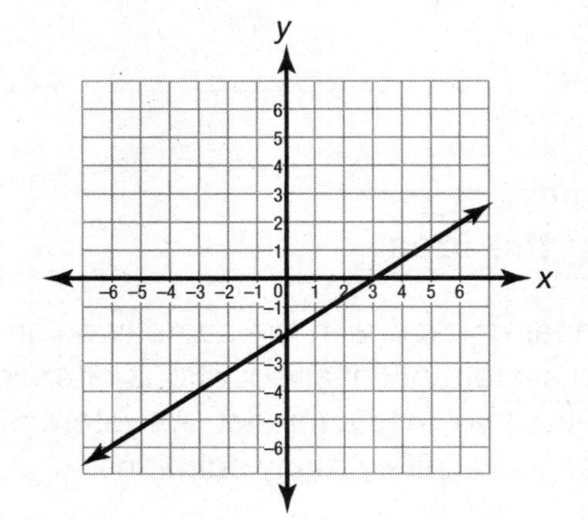

B.

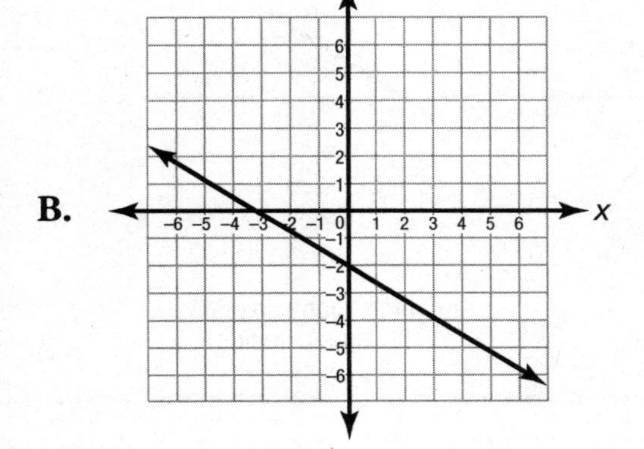

D.

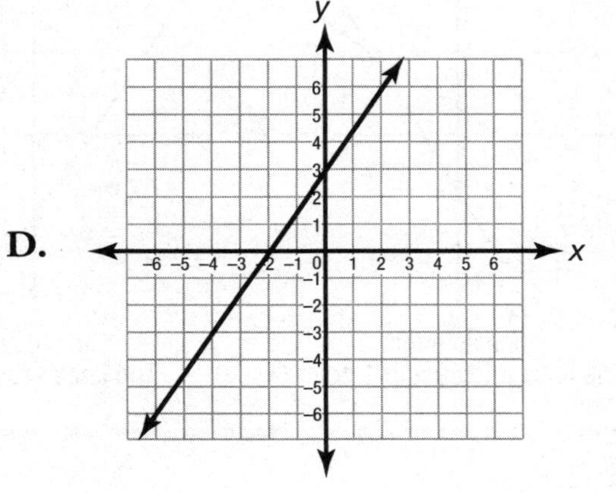

7. A line passes through points at (0, 7) and (−5, 0). What is the equation of the line in slope-intercept form?

Answer ______________

8. What is the equation of this line in slope-intercept form?

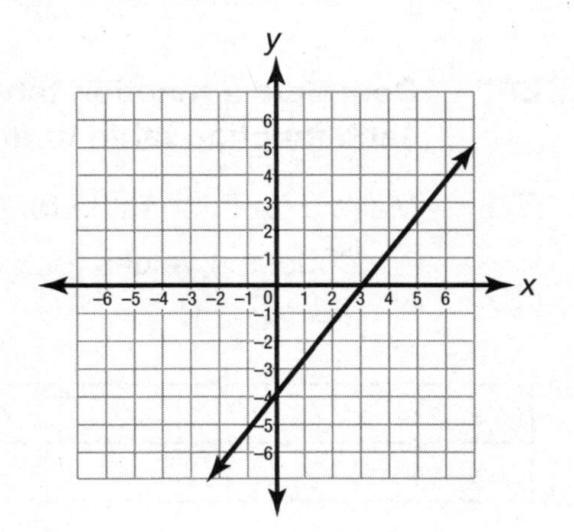

Answer ______________

Lesson

26 Systems of Linear Equations

Getting the Idea

Two linear equations in the same two variables form a **system of linear equations.** To solve a system of linear equations graphically, graph each equation and identify any points where the two lines intersect. A system of linear equations may have one solution, no solution, or infinitely many solutions.

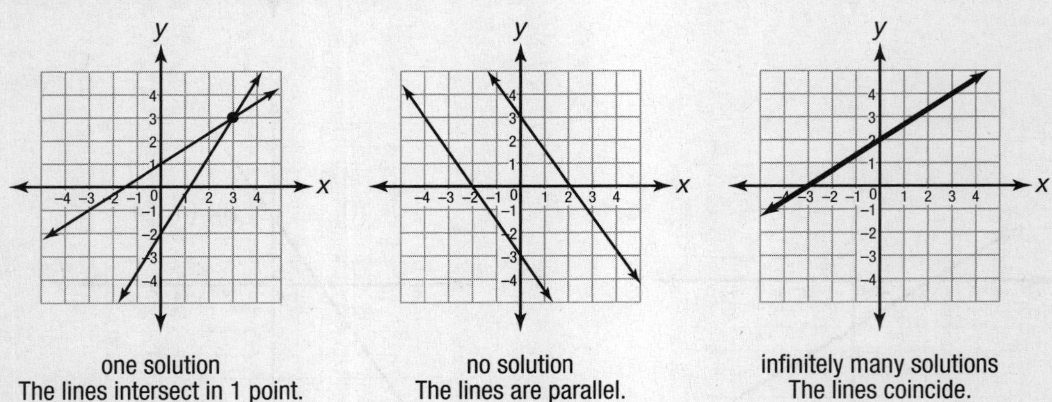

one solution
The lines intersect in 1 point.

no solution
The lines are parallel.

infinitely many solutions
The lines coincide.

EXAMPLE 1

Graph the system of linear equations shown below.

$y = 2x - 1$

$y = -x + 4$

STRATEGY **Complete a function table for each equation. Use the ordered pairs from each function table to graph the equations.**

STEP 1 Make a function table for each equation.

Choose 4 values for x and enter them into each function table.

$y = 2x - 1$

x	−2	0	2	4
y				

$y = -x + 4$

x	−2	0	2	4
y				

STEP 2 Complete each function table.

Substitute each x-value into the equations to find each corresponding y-value.

$y = 2x - 1$

For $x = -2$, $y = (2 \times -2) - 1 = -5$.

For $x = 0$, $y = (2 \times 0) - 1 = -1$.

For $x = 2$, $y = (2 \times 2) - 1 = 3$.

For $x = 4$, $y = (2 \times 4) - 1 = 7$.

$y = -x + 4$

For $x = -2$, $y = -(-2) + 4 = 6$

For $x = 0$, $y = -(0) + 4 = 4$

For $x = 2$, $y = -2 + 4 = 2$

For $x = 4$, $y = -4 + 4 = 0$

x	−2	0	2	4
y	−5	−1	3	7

x	−2	0	2	4
y	6	4	2	0

STEP 3 Write each x-value and corresponding y-value as an ordered pair for both equations.

$y = 2x - 1$

$(-2, -5), (0, -1), (2, 3), (4, 7)$

$y = -x + 4$

$(-2, 6), (0, 4), (2, 2), (4, 0)$

STEP 4 Graph the ordered pairs for the first equation on a coordinate grid and draw a line through them. Repeat this for the second equation.

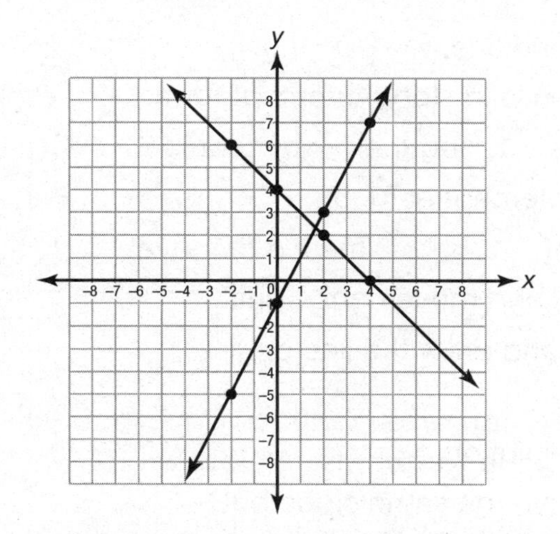

SOLUTION **The graph of the system of equations is shown in Step 4.**

EXAMPLE 2

Solve this system of linear equations by graphing.

$y = -x + 3$

$y = -x + 6$

STRATEGY **Graph the lines and determine the solution.**

STEP 1 Graph the first line, $y = -x + 3$.

The equation is in slope-intercept form.

The slope is -1, and the y-intercept is 3.

Plot the y-intercept at (0, 3).

The slope of -1, or $-\frac{1}{1}$, means to move down 1 unit and move right 1 unit.

Plot points and draw the line.

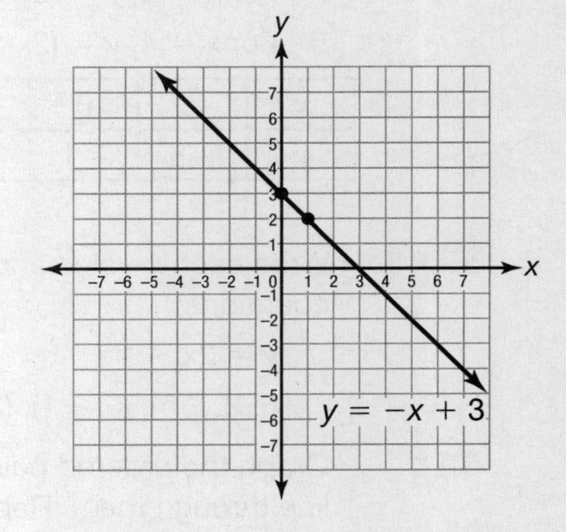

STEP 2 Graph the second line, $y = -x + 6$.

The equation is in slope-intercept form.

The slope is -1, and the y-intercept is 6.

Plot the y-intercept at (0, 6).

The slope of -1, or $-\frac{1}{1}$, means to move down 1 unit and move right 1 unit.

Plot points and draw the line on the same graph.

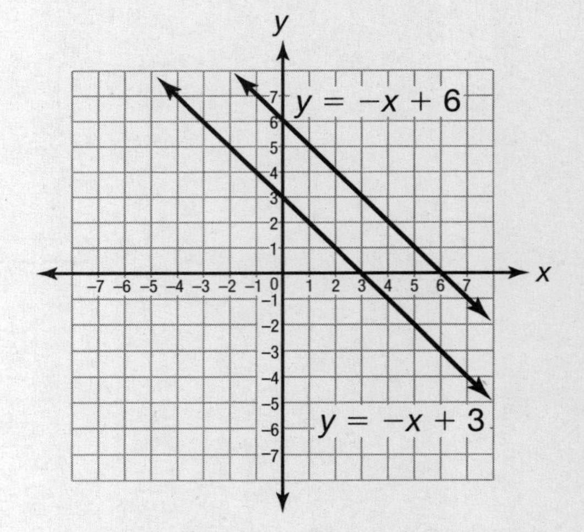

STEP 3 Determine the solution.

The lines have the same slope but different y-intercepts. The lines are parallel. Since the lines never intersect, there is no solution.

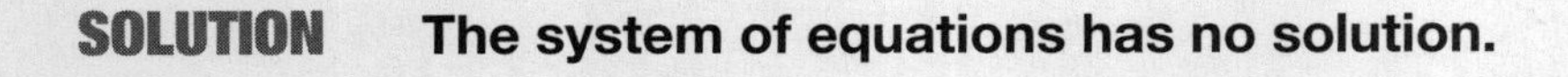

SOLUTION **The system of equations has no solution.**

EXAMPLE 3

Solve this system of linear equations by graphing.

$y = 2x + 3$

$y = 3x$

STRATEGY **Complete a function table for each equation. Use the ordered pairs from each function table to graph the equations, and then see where the lines intersect.**

STEP 1 Make a function table for each equation.

Choose 4 values for x and enter them into each function table.

$y = 2x + 3$

x	1	2	3	4
y				

$y = 3x$

x	1	2	3	4
y				

STEP 2 Complete each function table.

Substitute each x-value into the equations to find each corresponding y-value.

$y = 2x + 3$

For $x = -2$, $y = (2 \times -2) + 3 = -1$.

For $x = 0$, $y = (2 \times 0) + 3 = 3$.

For $x = 2$, $y = (2 \times 2) + 3 = 7$.

For $x = 4$, $y = (2 \times 4) + 3 = 11$.

$y = 3x$

For $x = -2$, $y = 3 \times -2 = -6$.

For $x = 0$, $y = 3 \times 0 = 0$.

For $x = 2$, $y = 3 \times 2 = 6$.

For $x = 4$, $y = 3 \times 4 = 12$.

x	1	2	3	4
y	5	7	9	11

x	1	2	3	4
y	3	6	9	12

STEP 3 Write each x-value and corresponding y-value as an ordered pair for both equations.

$y = 2x + 3$

(1, 5), (2, 7), (3, 9), (4, 11)

$y = 3x$

(1, 3), (2, 6), (3, 9), (4, 12)

STEP 4 Graph the ordered pairs for the first equation on a coordinate grid and draw a line through them. Repeat this for the second equation.

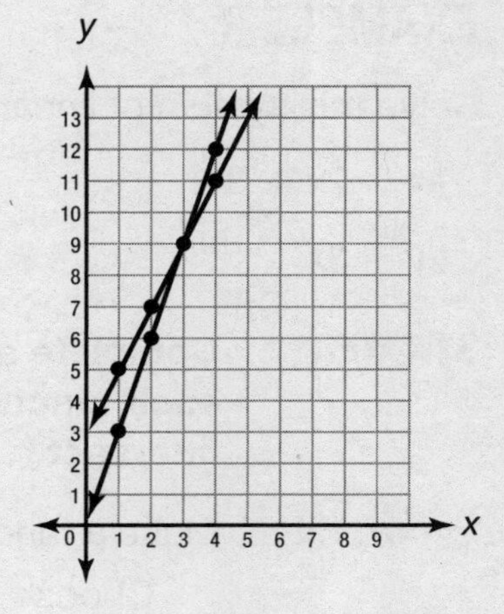

STEP 5 Locate the intersection point of the lines.

The lines intersect at the point with coordinates (3, 9). This ordered pair is the solution of the system of linear equations.

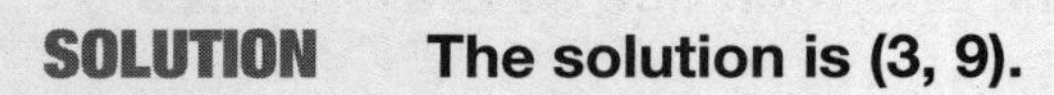

SOLUTION **The solution is (3, 9).**

COACHED EXAMPLE

Solve this system of linear equations by graphing:

$y = -x + 7$

$y = -\frac{3}{2}x + 9$

THINKING IT THROUGH

Make and complete a function table for each equation.

$y = -x + 7$

x	0	2	4	6
y	___	___	___	___

$y = -\frac{3}{2}x + 9$

x	0	2	4	6
y	___	___	___	___

Use the function tables to write the ordered pairs for both equations.

$y = -x + 7$

(0, ___), (2, ___),
(4, ___), (6, ___)

$y = -\frac{3}{2}x + 9$

(0, ___), (2, ___),
(4, ___), (6, ___)

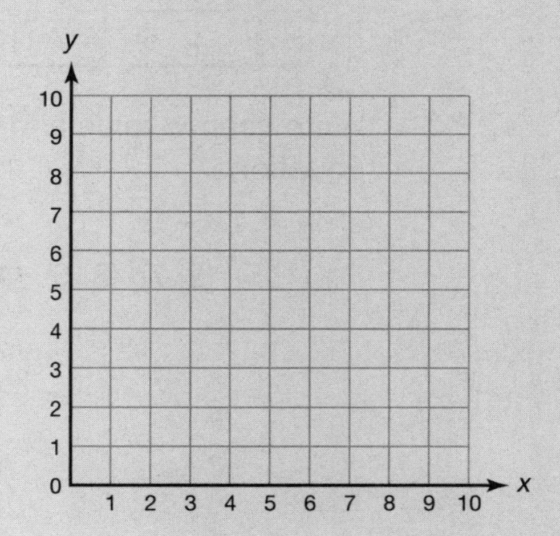

Graph both sets of ordered pairs.

The lines intersect at (___, ___).

The solution is __________.

Lesson Practice

Choose the correct answer.

1. What is the solution of the system of linear equations graphed below?

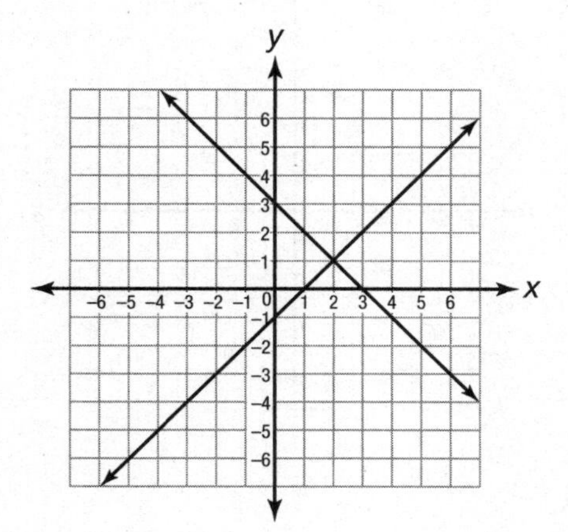

A. (1, 2)

B. (2, 1)

C. There is no solution.

D. There are infinitely many solutions.

2. What is the solution of the system of linear equations graphed below?

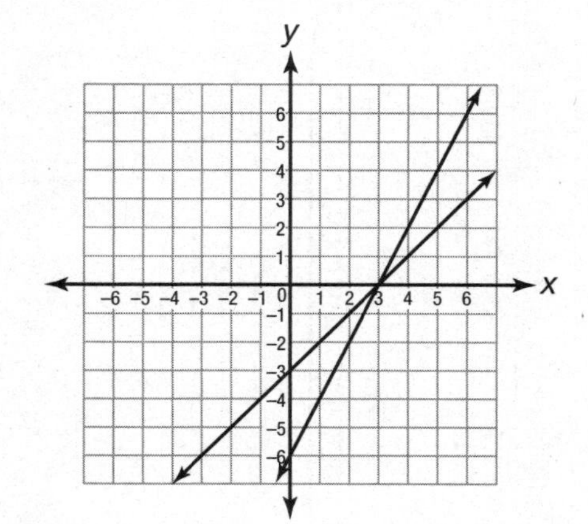

A. $(0, -3)$

B. $(0, -6)$

C. (3, 0)

D. There are infinitely many solutions.

3. What is the solution of this system of linear equations?

$y = -x + 4$

$y = 2x - 8$

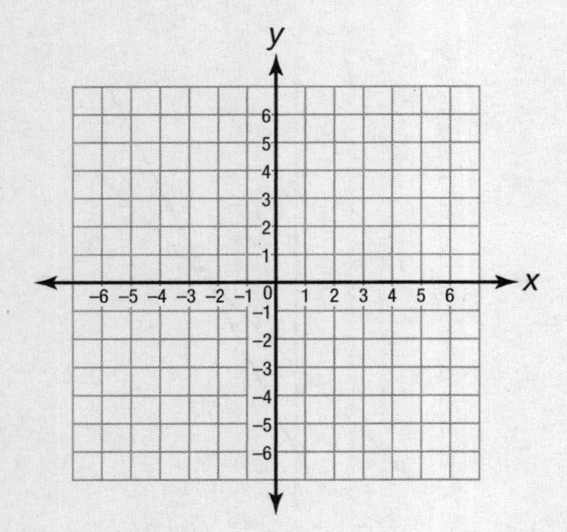

A. (4, 0)

B. (0, 4)

C. (−4, 0)

D. (0, −4)

4. Which system of linear equations has no solution?

A. $y = x - 2$
$y = -x - 3$

B. $y = x - 3$
$y = -x - 2$

C. $y = x - 2$
$y = x + 3$

D. $y = -x + 3$
$y = x + 2$

5. What is the solution of this system of linear equations?

$y = 2x + 3$

$y = 2x - 3$

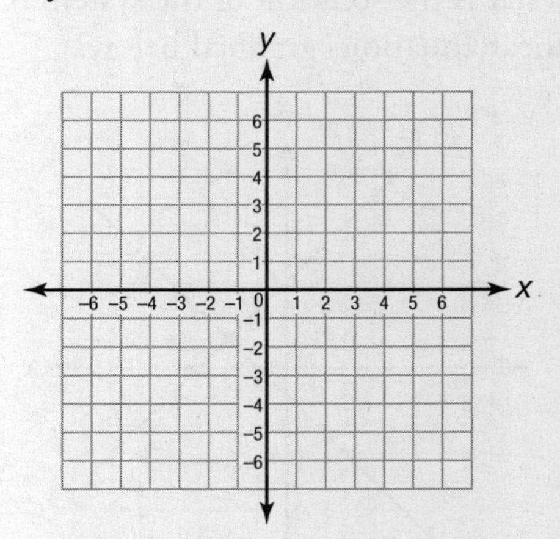

A. (−3, 2)

B. (0, 0)

C. There are infinitely many solutions.

D. There is no solution.

6. What is the solution of this system of linear equations?

$y = 4x + 6$

$y = 2x + 2$

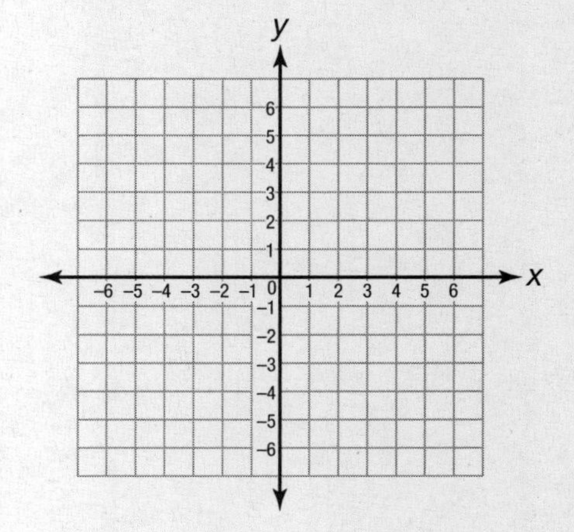

A. (2, 2)

B. (2, −2)

C. (−2, 2)

D. (−2, −2)

EXTENDED-RESPONSE QUESTION

7. Consider this system of linear equations.

$y = x + 1$ $\qquad$ $y = -2x - 2$

Part A Graph this system of linear equations.

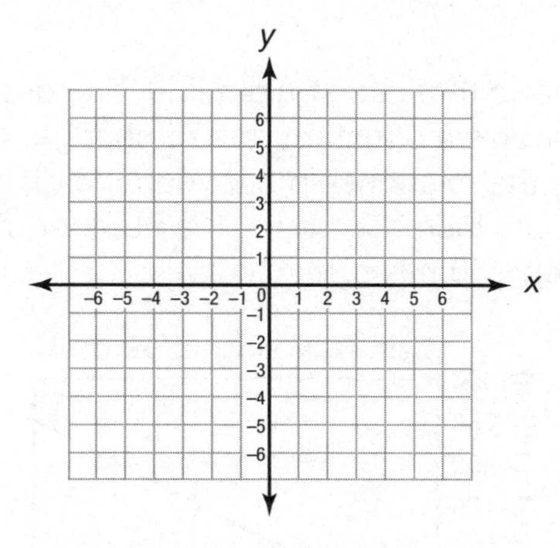

Part B What is the solution of the system? Explain.

____

____

____

____

Lesson

27 Graphs of Linear and Nonlinear Equations

Getting the Idea

A **linear equation** has a graph that is a straight line. The graph of a linear equation, such as the one shown below, shows a constant rate of change. A car is traveling at 50 miles per hour. The graph shows the relationship between the distance traveled and the time it takes. As the number of hours increases by 1, the number of miles increases by 50. After 5 hours, the car has traveled 250 miles.

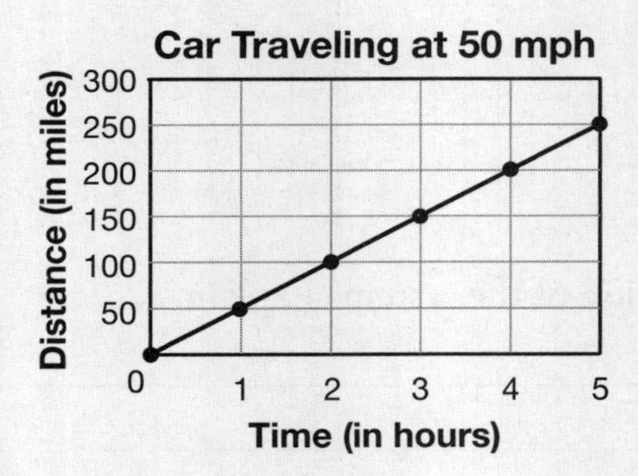

A **nonlinear equation** has a graph that is not a straight line. The graph of a nonlinear equation, such as the one shown below, shows a rate of change that is not constant. A golf ball is hit upward. The graph shows the relationship between the golf ball's height above the ground and the time it takes the golf ball to reach that height. As the number of seconds increases, the ball goes higher at a slower rate until it stops rising and starts to fall.

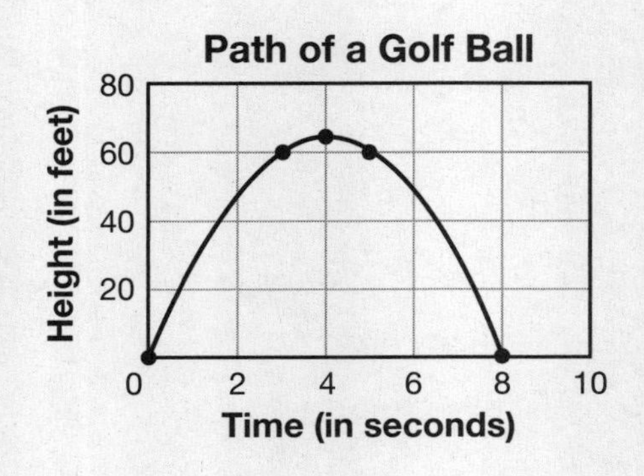

EXAMPLE 1

The table below shows some values for an equation. Plot the ordered pairs as points on a coordinate grid. Connect the points. Is the equation linear?

x	−2	−1	0	1	2
y	10	7	4	1	−2

STRATEGY **Identify and plot the ordered pairs.**

STEP 1 Write each x-value and corresponding y-value as an ordered pair.

(−2, 10), (−1, 7), (0, 4), (1, 1), (2, −2)

STEP 2 Graph the ordered pairs and connect the points.

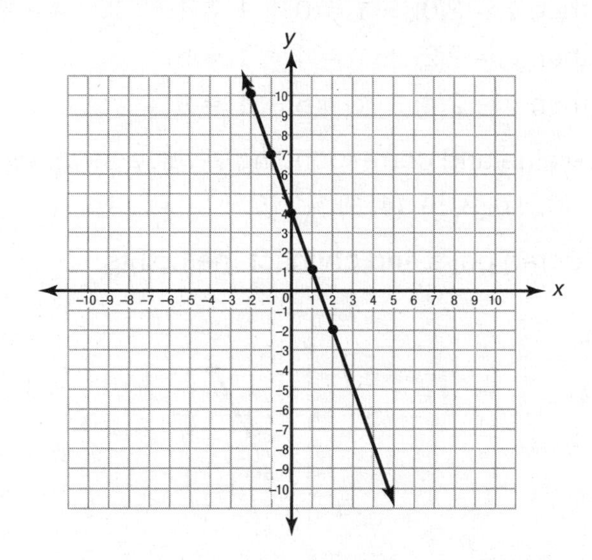

The points are all on the same line, so the graph is linear.

SOLUTION **The equation is linear.**

EXAMPLE 2

Complete the table of values for the equation $y = 2x + 1$. Then plot the ordered pairs as points on a coordinate grid and connect the points. Is the equation linear?

x	−2	0	2	4
y				

STRATEGY **Identify and plot the ordered pairs.**

STEP 1 Substitute each x-value into the equation $y = 2x + 1$ to find each corresponding y-value.

If $x = -2$, then $y = 2(-2) + 1 = -4 + 1 = -3$.

If $x = 0$, then $y = 2(0) + 1 = 0 + 1 = 1$.

If $x = 2$, then $y = 2(2) + 1 = 4 + 1 = 5$.

If $x = 4$, then $y = 2(4) + 1 = 8 + 1 = 9$.

STEP 2 Write each x-value and corresponding y-value as an ordered pair.

$(-2, -3), (0, 1), (2, 5), (4, 9)$

STEP 3 Graph the ordered pairs and connect the points.

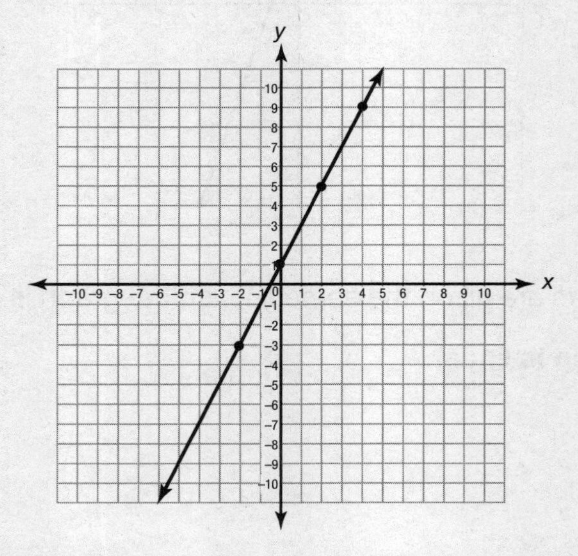

The points are all on the same line, so the graph is linear.

SOLUTION **The equation is linear.**

Both linear and nonlinear functions and their equations are used to model real-world data. Applications of linear and nonlinear functions include predicting costs and designing different kinds of products.

EXAMPLE 3

One graph below shows the amount of water accumulating in a barrel coming from a hose, and the other graph shows the path of a baseball after it is picked up and tossed into the air.

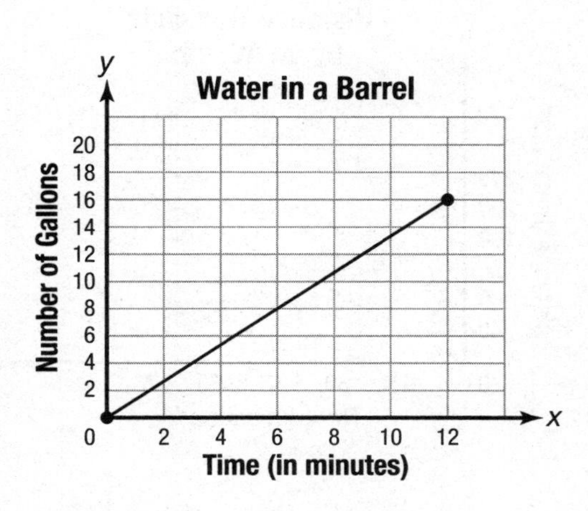

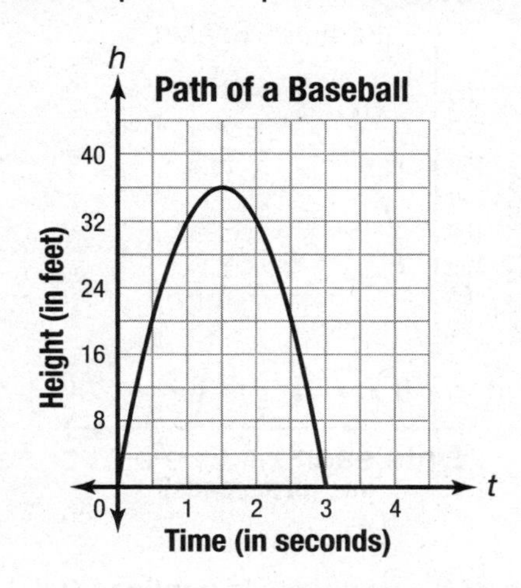

Which of the two functions is linear?

STRATEGY **Use the definitions of the graphs of linear and nonlinear functions.**

STEP 1 The graph of a linear function is a straight line. Does either graph show a straight line?

The water in a barrel graph shows a straight line, so the function is linear.

STEP 2 The graph of a nonlinear function is curved. Does either graph show a curve?

The path-of-a-baseball graph shows a curve, so the function is nonlinear.

SOLUTION **The water in a barrel function is linear.**

COACHED EXAMPLE

The graph on the left shows the path of a rocket when shot vertically into the air. The graph on the right shows the distance traveled by an athlete running at a constant speed of 3 meters per second for 90 seconds.

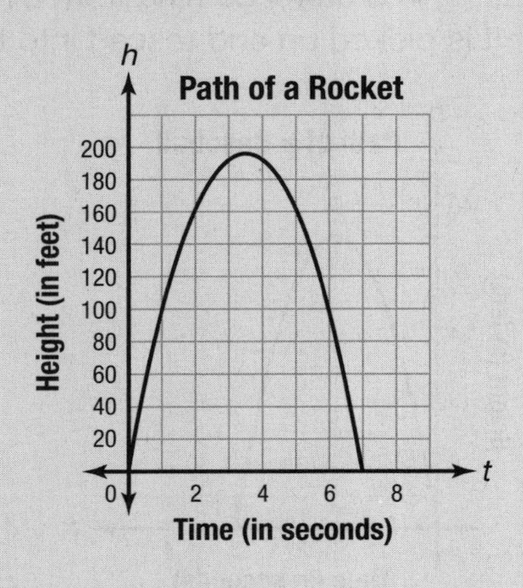

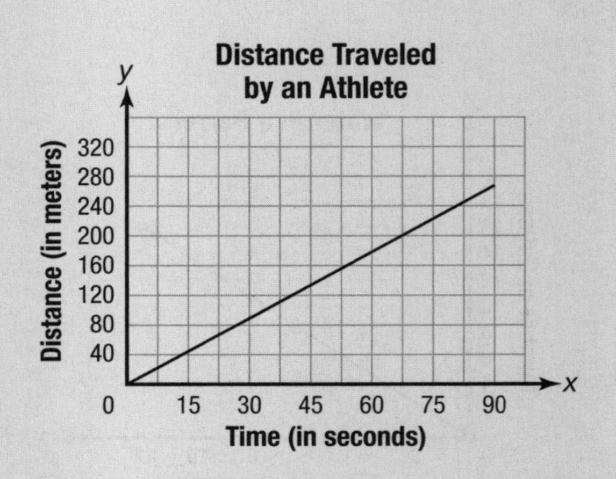

Which of the functions is nonlinear?

THINKING IT THROUGH

The graph of a linear function is a ___________ line.

The graph of a ____________________ function is a curve.

The path-of-a-rocket graph shows a ___________.

The distance-traveled-by-an-athlete graph shows a ____________________.

The ____________________ function is nonlinear.

Lesson Practice

Choose the correct answer.

1. Which of the following best describes the equation graphed below?

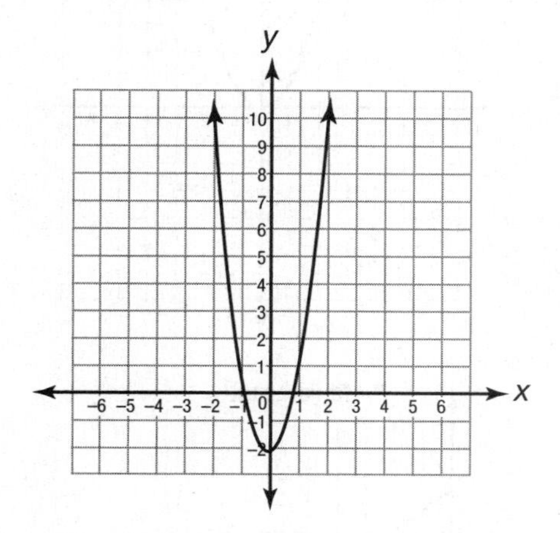

A. nonlinear because the graph represents a straight line

B. linear because the graph represents a curve

C. nonlinear because the graph represents a curve

D. linear because the graph represents a straight line

2. Which of the following is the graph of a linear equation?

A.

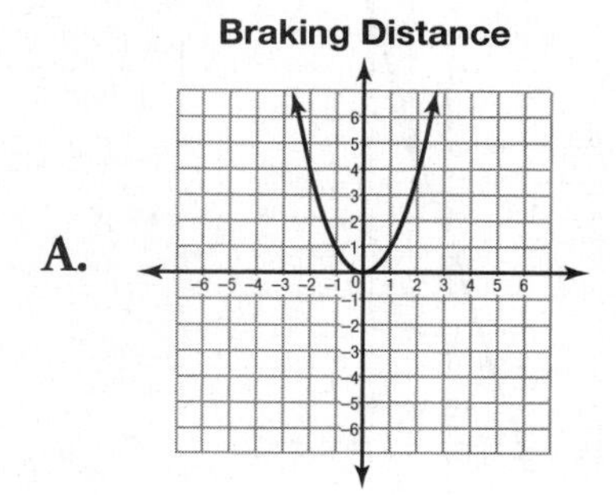

B.

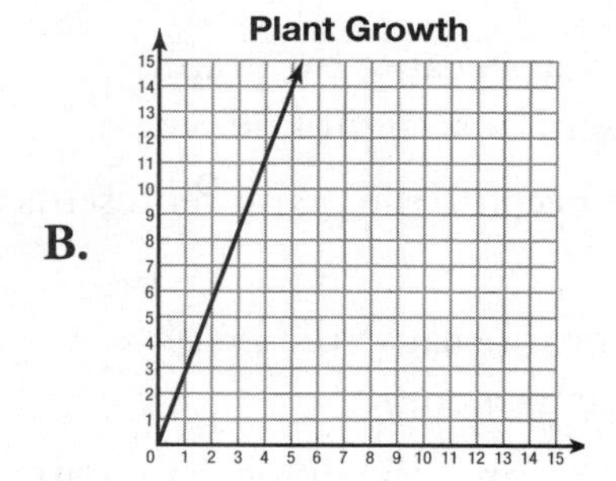

C.

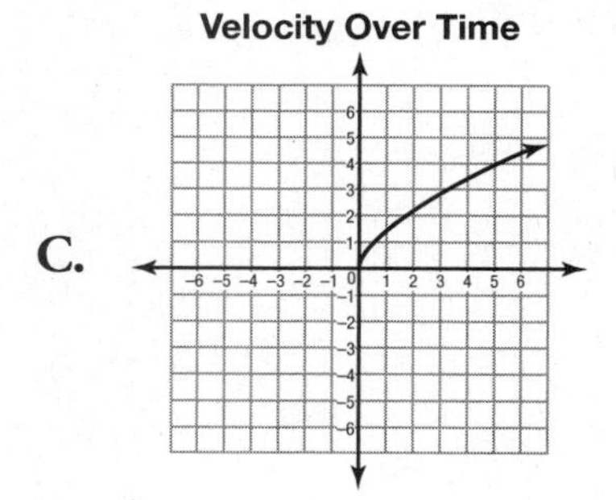

D.

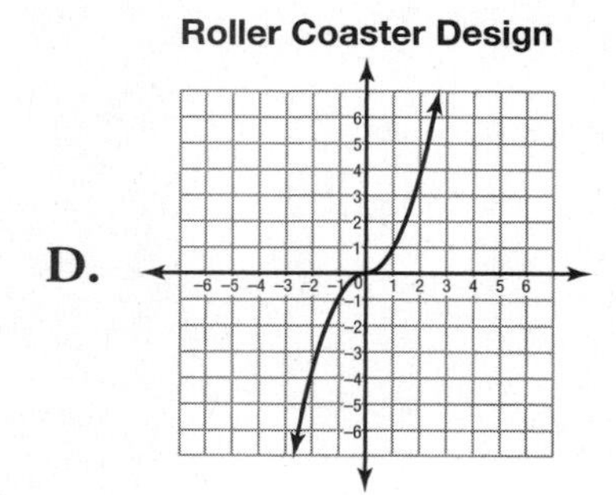

3. Which of the following best describes the equation graphed below?

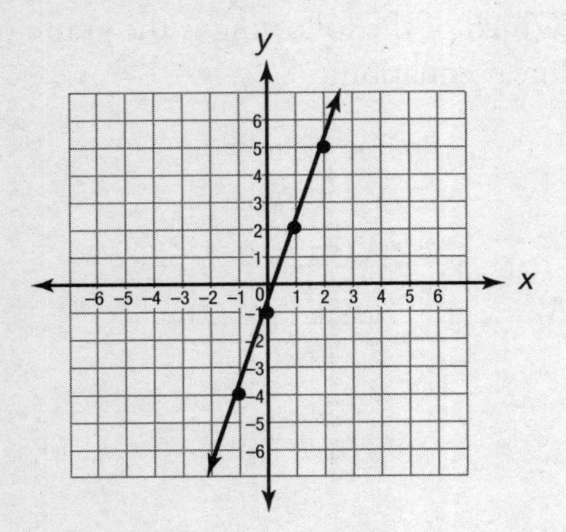

A. nonlinear because the graph represents a straight line

B. linear because the graph represents a curve

C. nonlinear because the graph represents a curve

D. linear because the graph represents a straight line

4. Which of the following is the graph of a nonlinear equation?

A.

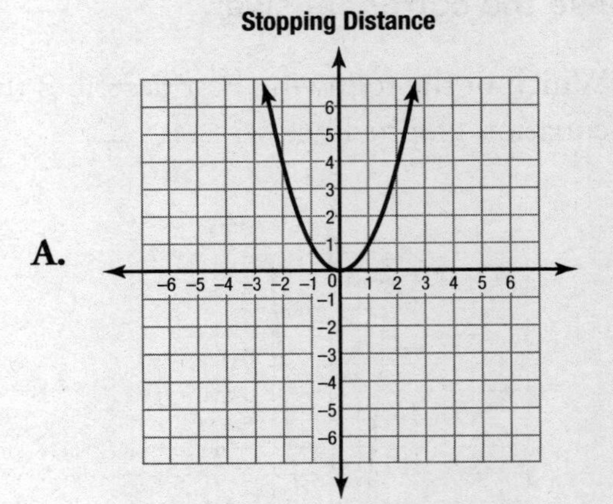

B.

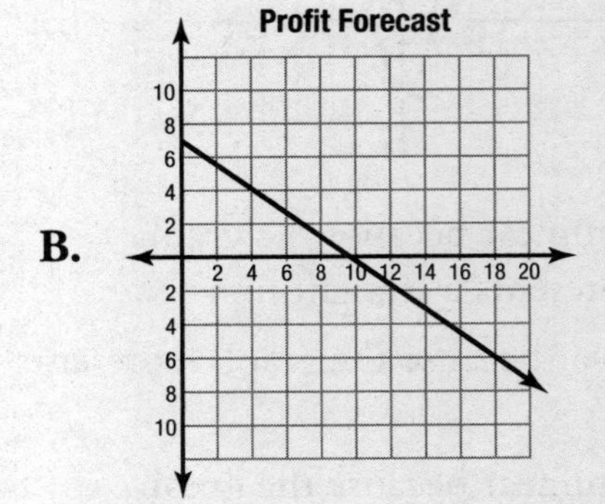

C.

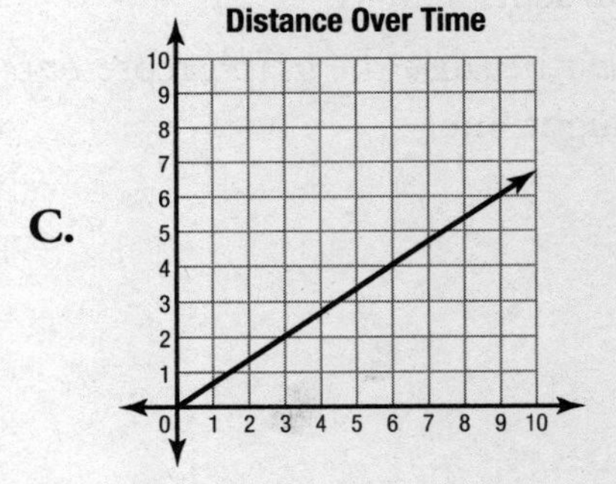

D.

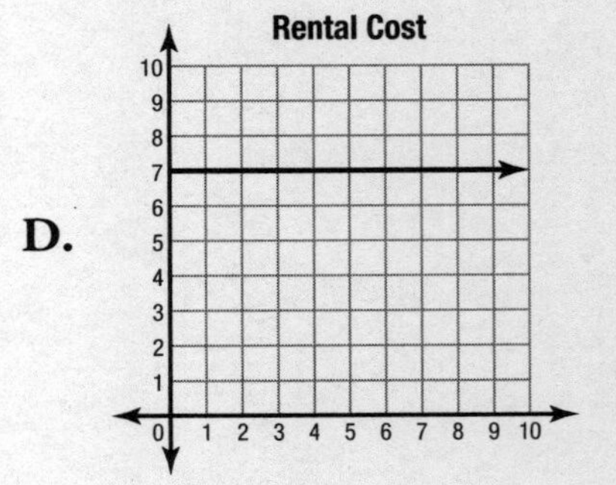

EXTENDED-RESPONSE QUESTION

5. This is a table of values for an equation.

x	0	1	2	3	4	5	6
y	5	0	−3	−4	−3	0	5

Part A Plot the ordered pairs on the coordinate grid below. Then connect the points.

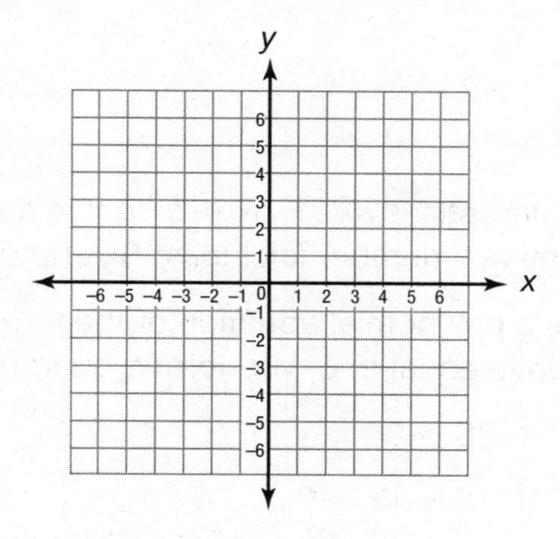

Part B Is the equation linear or nonlinear? Explain.

____

____

____

Lesson

28 Quadratic Equations

8.G.21

Getting the Idea

A **quadratic** is a polynomial with a variable raised to the second power. Here are some examples:

x^2 $\qquad$ $x^2 - 25$

$x^2 + 4x$ $\qquad$ $x^2 + 7x + 12$

The general form for a quadratic is $ax^2 + bx + c$. In this form, x is the variable, a is any real number except 0, b is any real number (including 0), and c is any real number (including 0).

A **quadratic equation** is a polynomial equation of the form $ax^2 + bx + c = 0$. By setting any of the quadratics above equal to 0, you form a quadratic equation.

$x^2 = 0$ $\qquad$ $x^2 - 25 = 0$

$x^2 + 4x = 0$ $\qquad$ $x^2 + 7x + 12 = 0$

EXAMPLE 1

Which of the following is a quadratic equation?

$\frac{x}{2} + 1 = 0$ $\qquad$ $x^2 + 1 = 0$

$2x + 1 = 0$ $\qquad$ $x^3 + 1 = 0$

STRATEGY **Look for an equation that has the variable raised to the second power.**

In $\frac{x}{2} + 1 = 0$, the variable x is raised to the first power.

In $2x + 1 = 0$, the variable x is raised to the first power.

In $x^2 + 1 = 0$, the variable x is raised to the second power.

In $x^3 + 1 = 0$, the variable x is raised to the third power.

SOLUTION **The quadratic equation is $x^2 + 1 = 0$.**

A **quadratic function** has the form $y = ax^2 + bx + c$. The graph of a quadratic function is a curve called a **parabola**. If the coefficient a of ax^2 is positive, the parabola opens upward. If the coefficient a of ax^2 is negative, the parabola opens downward. Here are two examples of parabolas.

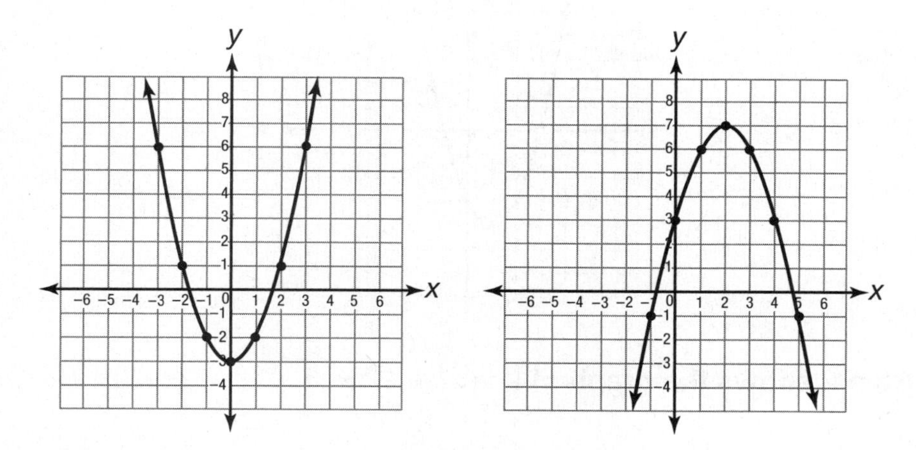

You can graph a quadratic function by substituting values for x to find corresponding values for y. Then plot the points and connect them with a smooth curve to form the parabola.

EXAMPLE 2

Graph the quadratic function $y = x^2 + 2x - 3$.

STRATEGY **Create a table of values by substituting values for x. Then plot the points and connect them with a curve.**

STEP 1 Create a table of values.

x	$y = x^2 + 2x - 3$	y	(x, y)
−4	$y = (-4)^2 + 2(-4) - 3 = 5$	5	(−4, 5)
−3	$y = (-3)^2 + 2(-3) - 3 = 0$	0	(−3, 0)
−2	$y = (-2)^2 + 2(-2) - 3 = -3$	−3	(−2, −3)
−1	$y = (-1)^2 + 2(-1) - 3 = -4$	−4	(−1, −4)
0	$y = (0)^2 + 2(0) - 3 = -3$	−3	(0, −3)
1	$y = (1)^2 + 2(1) - 3 = 0$	0	(1, 0)
2	$y = (2)^2 + 2(2) - 3 = 5$	5	(2, 5)

STEP 2 Plot the points and connect them with a curve to form a parabola.

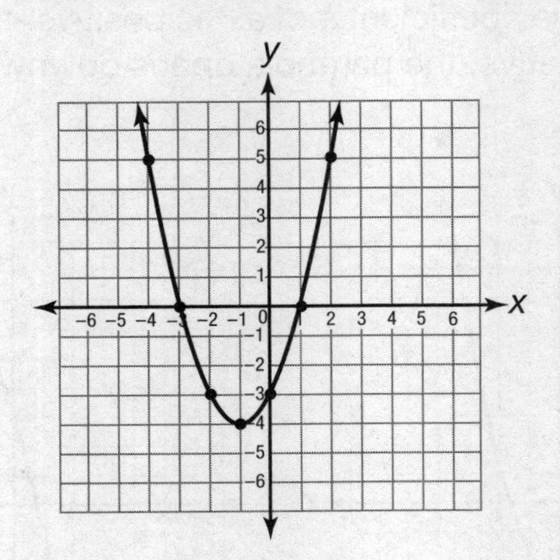

SOLUTION **Step 2 shows the graph of $y = x^2 + 2x - 3$.**

You can tell whether a table of x- and y-values represents a quadratic function by looking for patterns in the x- and y-values. If the x-values are consecutive integers and the second differences of consecutive y-values are all 2, then the table represents a quadratic function. In the table in Example 2, the x-values are consecutive integers. The y-values from the table and their first differences and second differences are shown below.

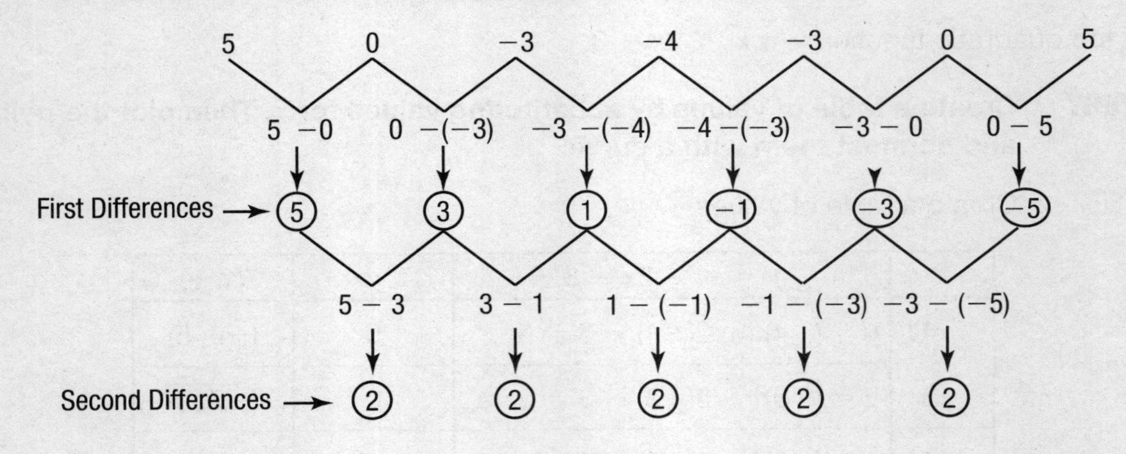

The second differences are all 2, indicating that the table represents a quadratic function.

EXAMPLE 3

Does this table represent a quadratic function?

x	−2	−1	0	1	2
y	28	18	10	4	0

STRATEGY **If the x-values are consecutive, find the second differences of the y-values.**

STEP 1 Are the x-values consecutive integers?

−2, −1, 0, 1, 2 are consecutive integers.

STEP 2 Find the first differences in the y-values.

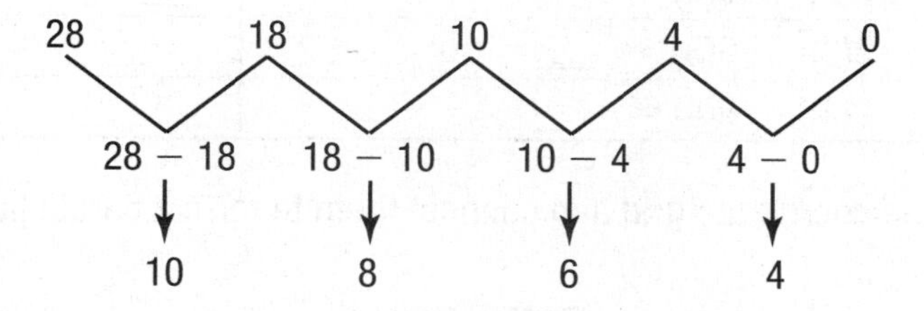

The first differences are 10, 8, 6, and 4.

STEP 3 Find the second differences.

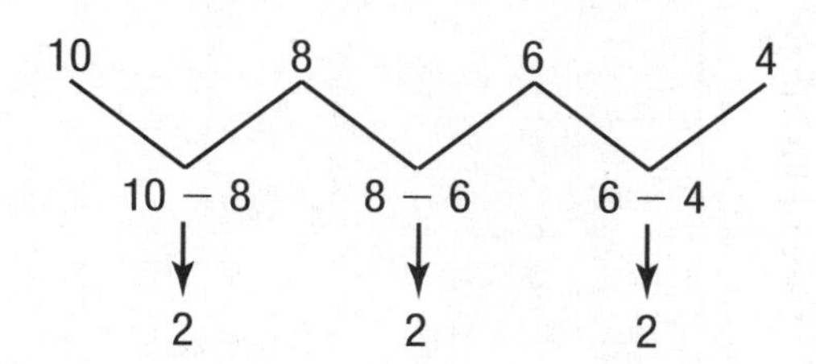

The second differences are all 2.

SOLUTION **The table represents a quadratic function.**

COACHED EXAMPLE

Matt tosses a coin into the air. Its path is given by $h = -2t^2 + 8t$, where h is the height in inches of the coin, and t is the time in seconds. Graph this situation.

THINKING IT THROUGH

Make a table of values.

t	$h = -2t^2 + 8t$	h	(t, h)
0	$h = -2(0)^2 + 8(0) = 0$	0	(0, 0)
1	$h = -2(1)^2 - 8(1) =$ ____	____	(__, __)
2	$h = -2(2)^2 + 8(2) =$ ____	____	(__, __)
3	$h = -2(3)^2 + 8(3) =$ ____	____	(__, __)
4	$h = -2(4)^2 + 8(4) =$ ____	____	(__, __)

Plot the points on the coordinate grid and connect them to form a parabola.

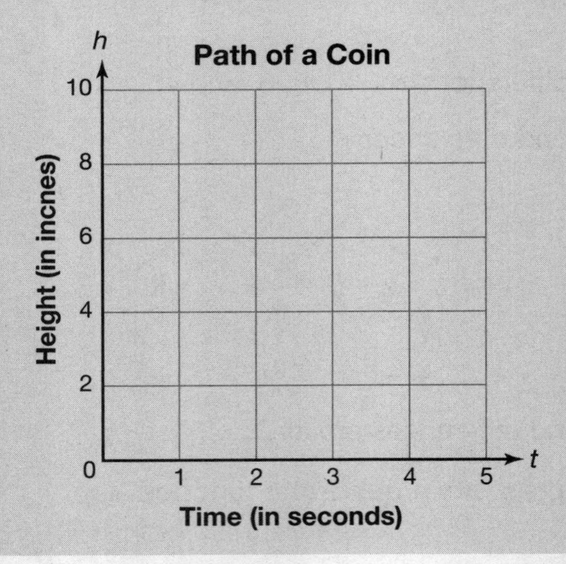

Lesson Practice

Choose the correct answer.

1. Which of the following is the graph of a quadratic function?

A.

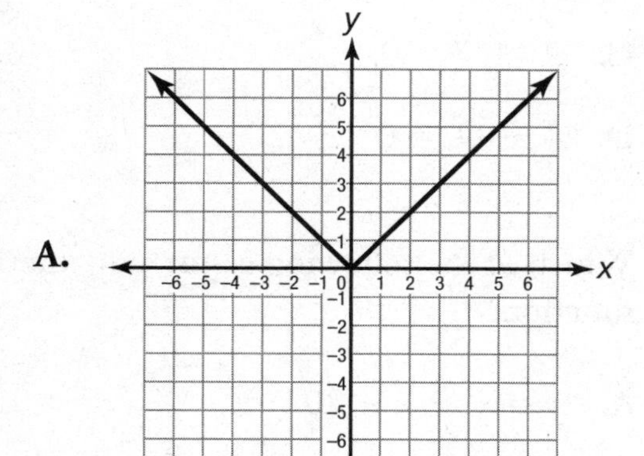

B.

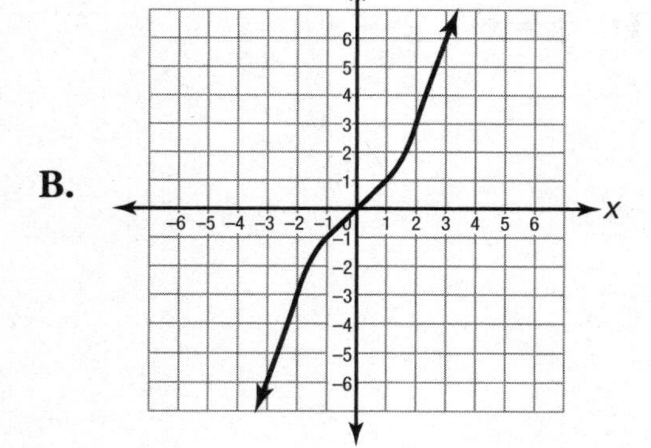

C.

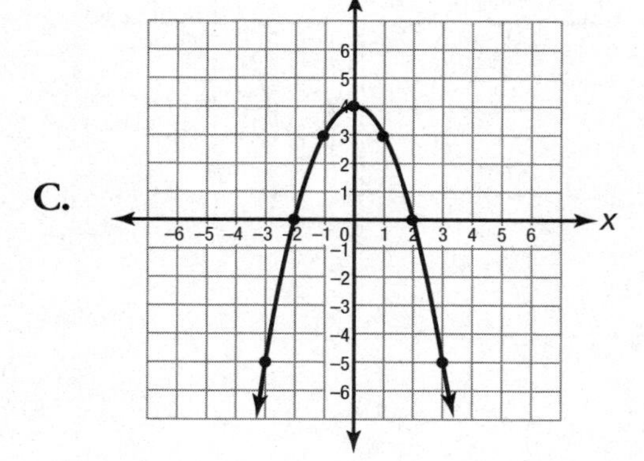

D.

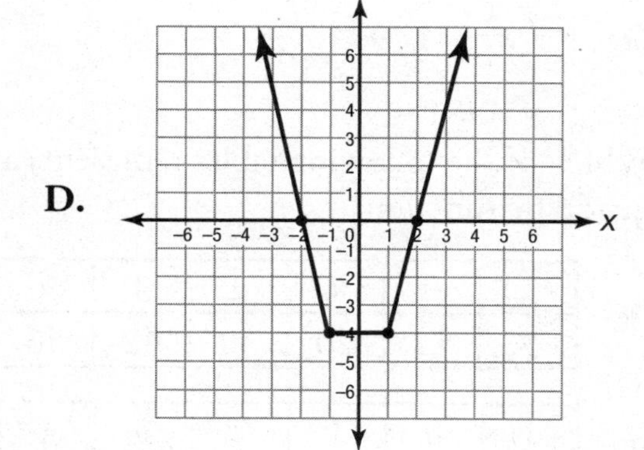

2. Which quadratic function has this graph?

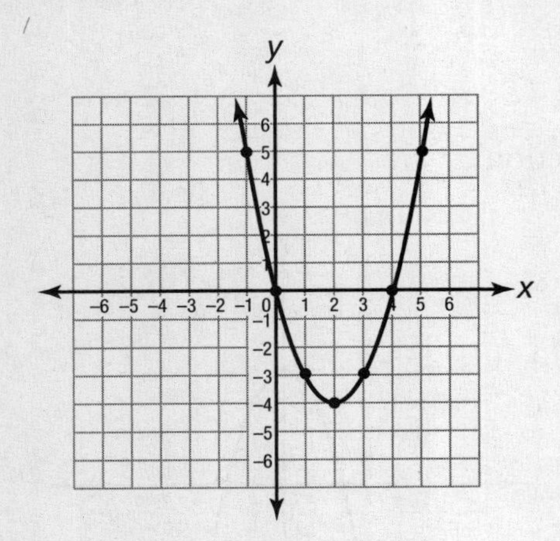

A. $y = x^2 - 4$

B. $y = x^2 - 4x$

C. $y = x^2 + 4x$

D. $y = x^2 - 4x + 4$

3. Which of the following tables represents a quadratic function?

A.

x	0	1	2	3	4
y	12	20	30	42	56

B.

x	0	1	2	3	4
y	12	30	28	36	44

C.

x	0	1	2	3	4
y	0	1	4	8	12

D.

x	0	1	2	3	4
y	15	30	45	60	75

4. Which of the following is a quadratic equation?

A. $x^2 - 8x + 15 = 0$

B. $x^3 + 8x + 2 = 0$

C. $2x + 7 = 0$

D. $\frac{x}{4} + 11 = 0$

5. Which of the following is **not** a quadratic function?

A. $y = x^2 - x - 20$

B. $y = x^2 - 36$

C. $y = x^2 + 36x$

D. $y = x^4 + x^3 + x^2$

EXTENDED-RESPONSE QUESTION

6. A baseball is thrown into the air. Its path is given by $h = -16t^2 + 48t$, where h is the height in feet and t is the time in seconds.

Part A Complete the table below.

t	$h = -16t^2 + 48t$	h	(t, h)
0	$h = -16(0)^2 + 48(0) =$		(____, ____)
1			(____, ____)
1.5			(____, ____)
2			(____, ____)
3			(____, ____)

Part B Graph the function.

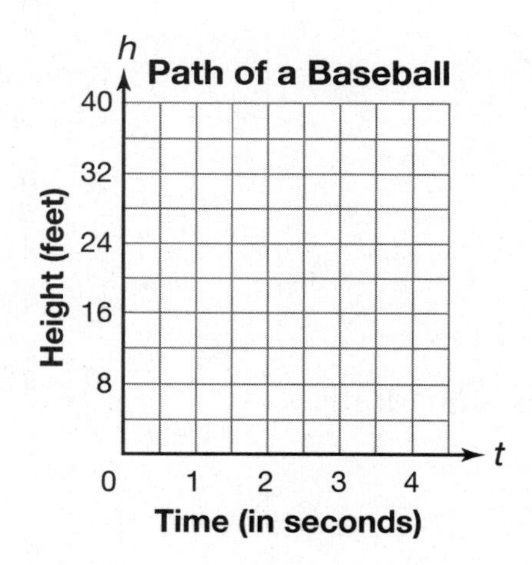

3 Review

1 In the diagram below, lines *a*, *b*, and *c* intersect. Which angle must be congruent to ∠3?

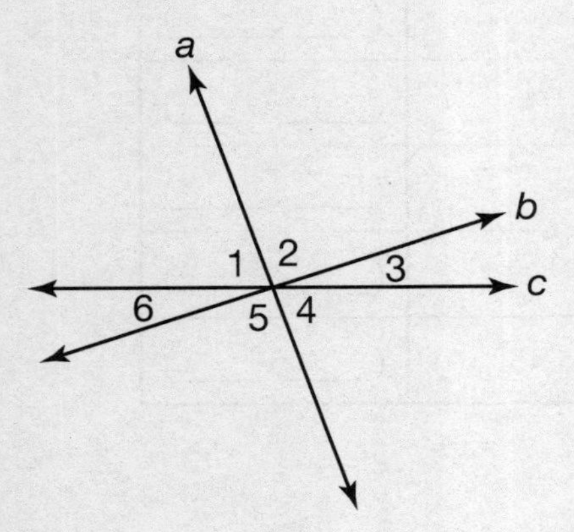

A ∠1

B ∠4

C ∠5

D ∠6

2 Which of the following is a quadratic equation?

A $3x - 8 = 0$

B $x^3 - 8 = 0$

C $x^2 - 8 = 0$

D $\frac{2}{x} - 8 = 0$

3 In the diagram below, line *j* is parallel to line *k*, and line *l* is a transversal.

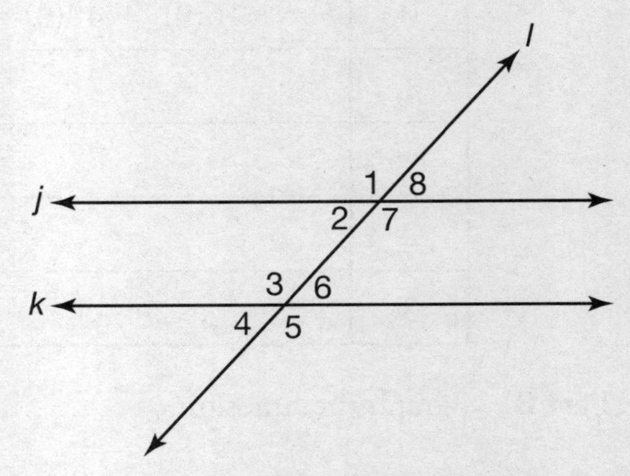

If m∠3 = 132°, what is the measure of ∠7?

A 48°

B 123°

C 132°

D 168°

4 Which of the following **best** describes the transformation of triangle MNO to M′N′O′?

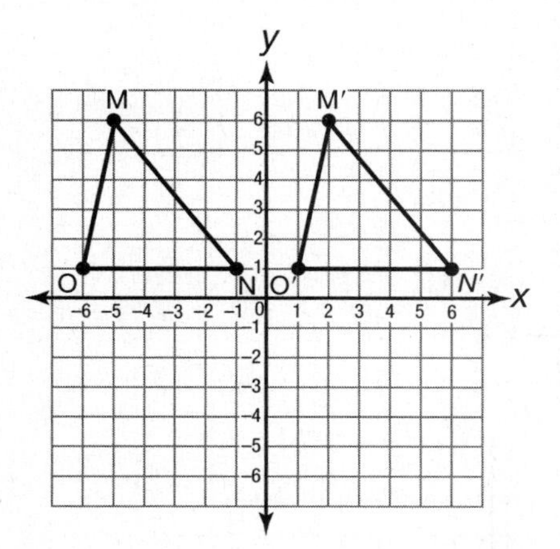

A translation of 7 units to the left

B translation of 7 units to the right

C reflection over the y-axis

D 90° clockwise rotation about the origin

5 A line passes through points at (−4, 0) and (0, 1) and has a slope of $\frac{1}{4}$. What is the equation of the line in slope-intercept form?

A $y = \frac{1}{4}x - 1$

B $y = \frac{1}{4}x + 1$

C $y = \frac{1}{4}x - 4$

D $y = \frac{1}{4}x + 4$

6 Which image shows a dilation of triangle ABC with a scale factor of 2?

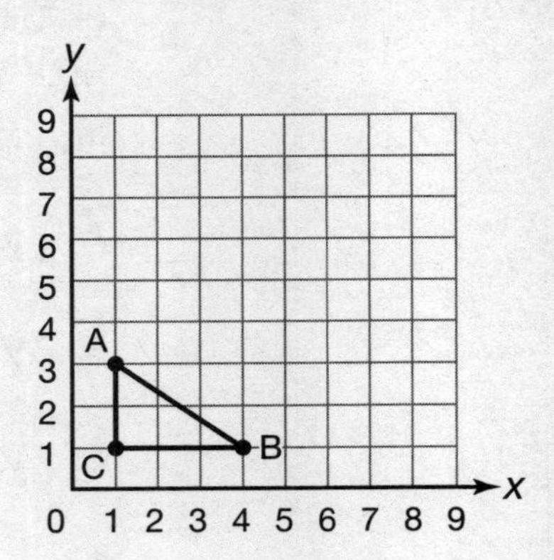

A

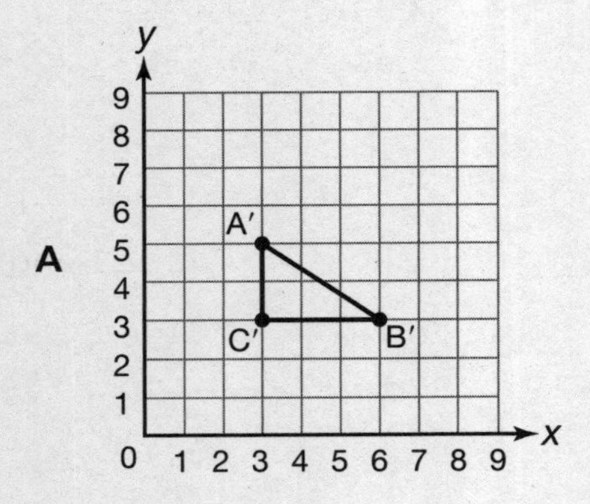

C

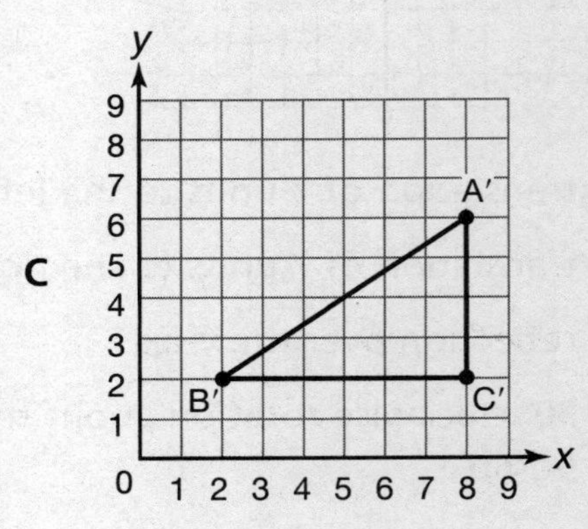

B

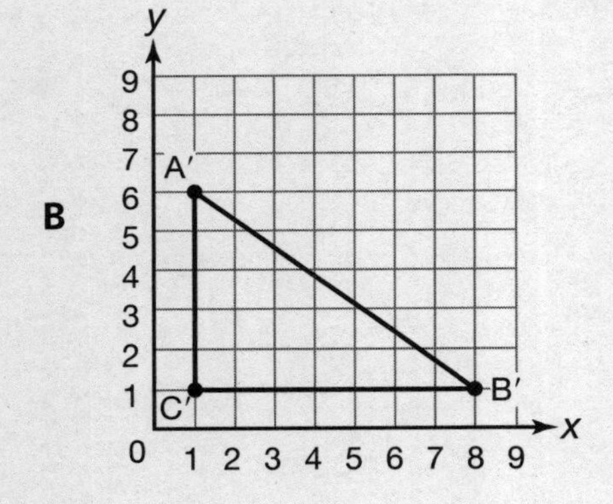

D

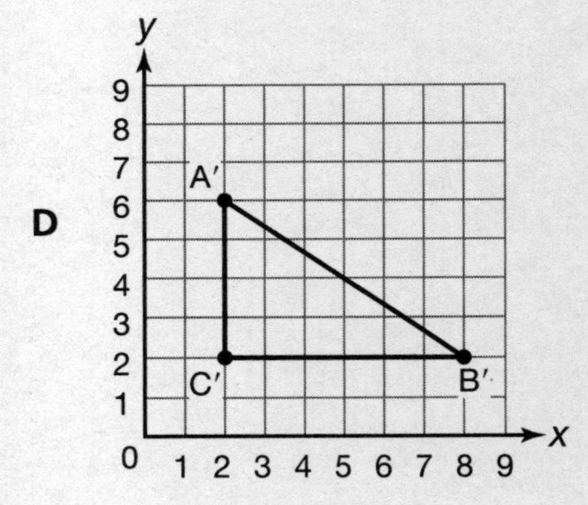

7 What is the solution of this system of equations?

$y = -x + 5$

$y = 2x - 4$

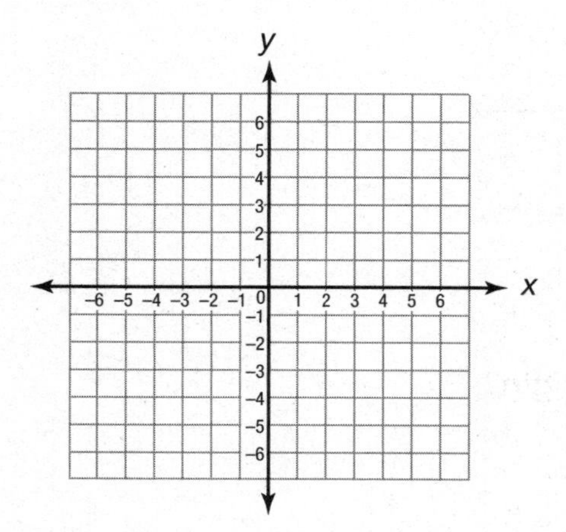

A $(-3, -2)$ C $(-2, -3)$

B $(3, 2)$ D $(2, 3)$

8 In the diagram below, which pair of angles is supplementary?

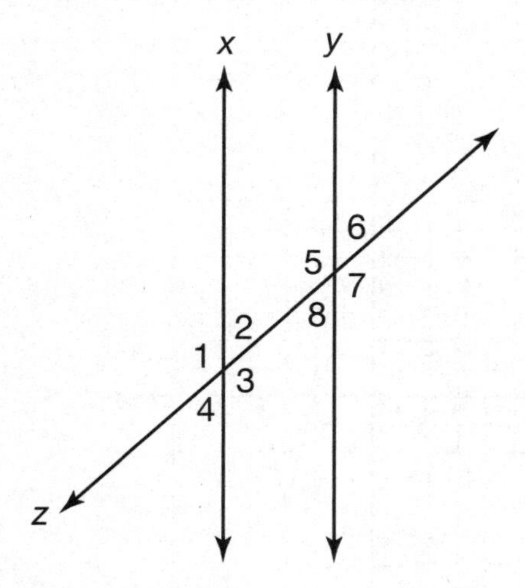

A $\angle 1$ and $\angle 5$

B $\angle 2$ and $\angle 5$

C $\angle 3$ and $\angle 5$

D $\angle 7$ and $\angle 5$

9 In the diagram below, line *r* is parallel to line *s*, and tranversal *t* intersects both lines.

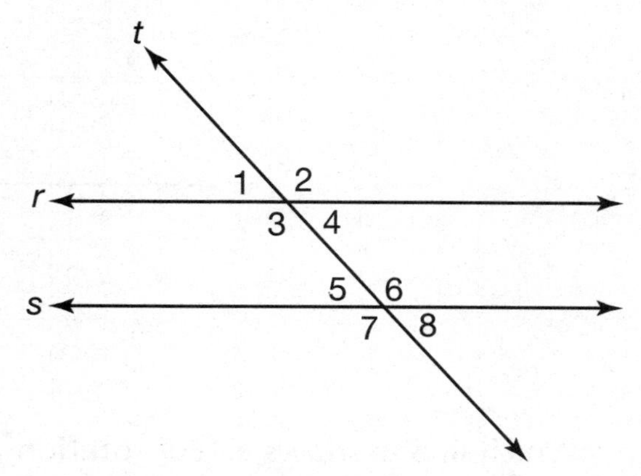

Which pair of angles are alternate exterior angles?

A $\angle 1$ and $\angle 7$ C $\angle 1$ and $\angle 8$

B $\angle 1$ and $\angle 6$ D $\angle 2$ and $\angle 8$

10 In the diagram below, lines *x* and *y* intersect, and $m\angle 1 = 34°$.

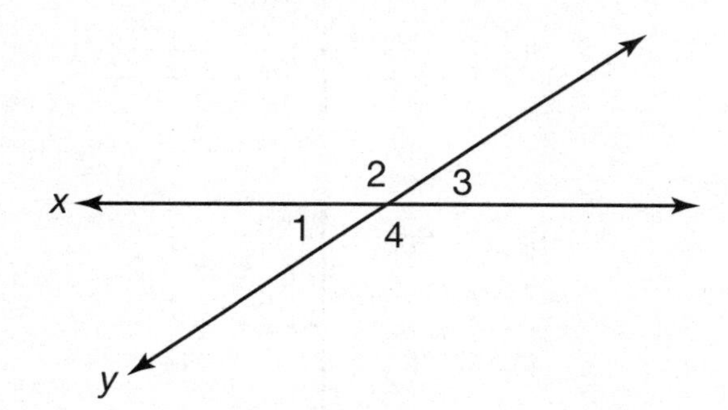

What are the measures of the other angles?

A $m\angle 2 = 34°$; $m\angle 3 = 34°$; $m\angle 4 = 34°$

B $m\angle 2 = 34°$; $m\angle 3 = 146°$; $m\angle 4 = 146°$

C $m\angle 2 = 146°$; $m\angle 3 = 146°$; $m\angle 4 = 34°$

D $m\angle 2 = 146°$; $m\angle 3 = 34°$; $m\angle 4 = 146°$

11 A triangle is plotted on the coordinate plane below.

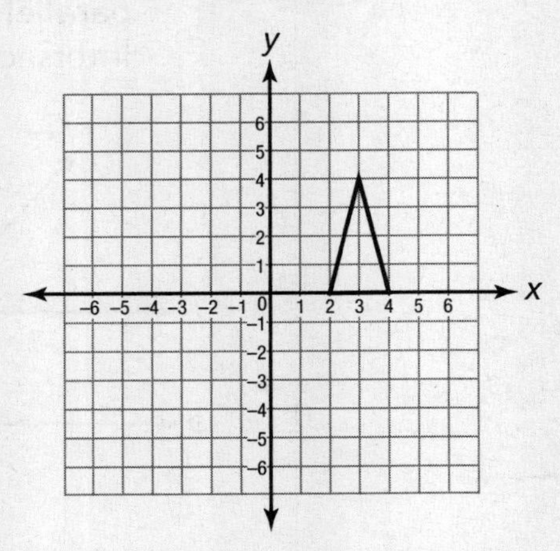

Which image shows a 180° rotation about the origin?

A
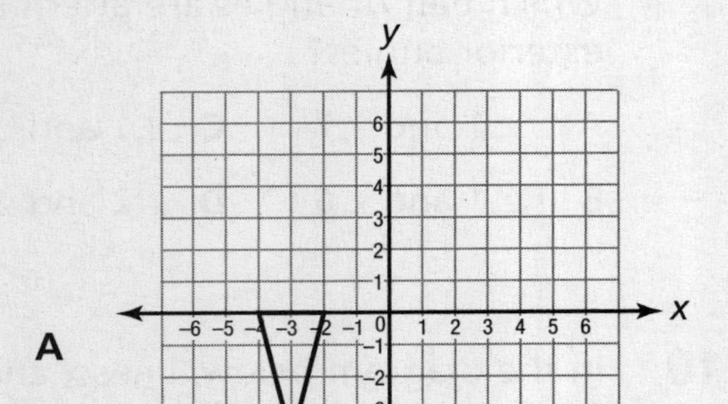

C
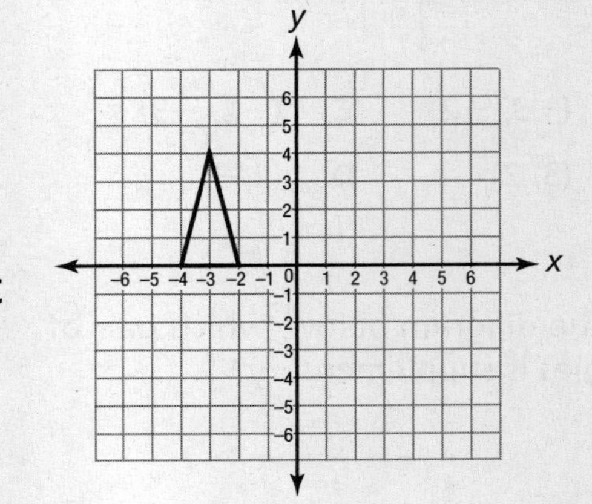

B

D
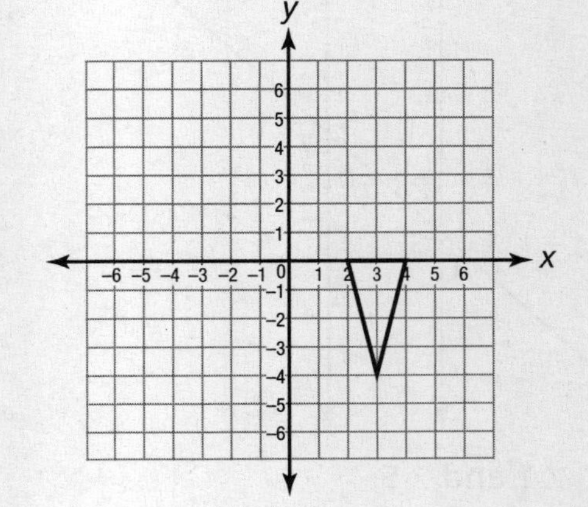

12 What are the coordinates of the image of triangle MNO after a translation of 3 units to the left and 2 units up?

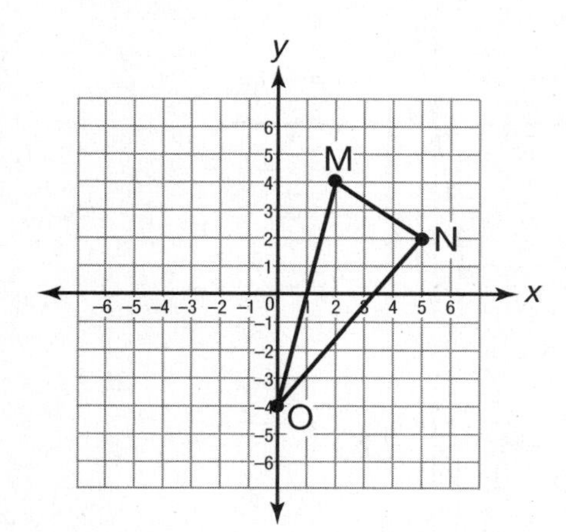

A M′(−1, 6), N′(2, 4), O′(−3, −2)

B M′(5, 2), N′(7, 0), O′(0, −6)

C M′(−1, 2), N′(2, 0), O′(−3, −6)

D M′(4, 1), N′(7, −1), O′(2, −7)

13 The area of triangle XYZ is 18 square feet. Under which transformation could the area of the image, triangle X′Y′Z′, be **greater** than 18 square feet?

A dilation

B reflection

C rotation

D translation

14 What is the slope of this line?

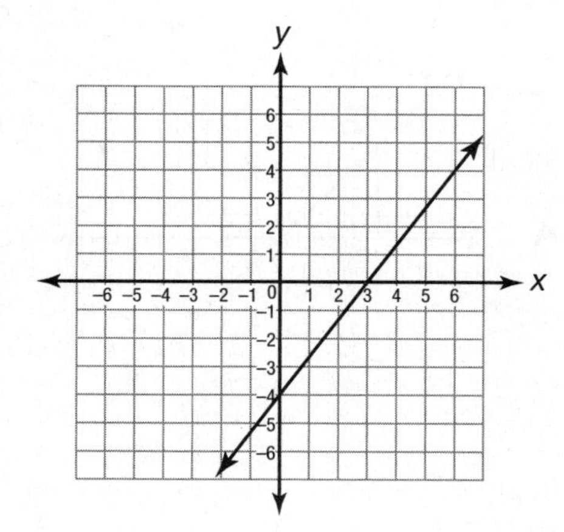

A $-\frac{4}{3}$

B $-\frac{3}{4}$

C $\frac{3}{4}$

D $\frac{4}{3}$

15 What is the *y*-intercept of this line?

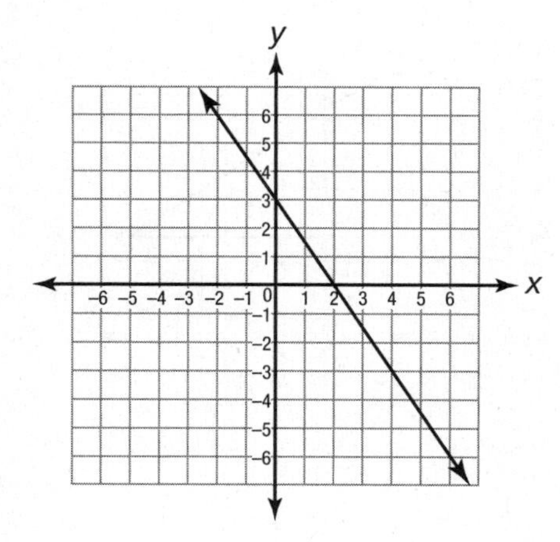

A −3

B 2

C 0

D 3

16 Which of the following is the graph of a nonlinear equation?

A
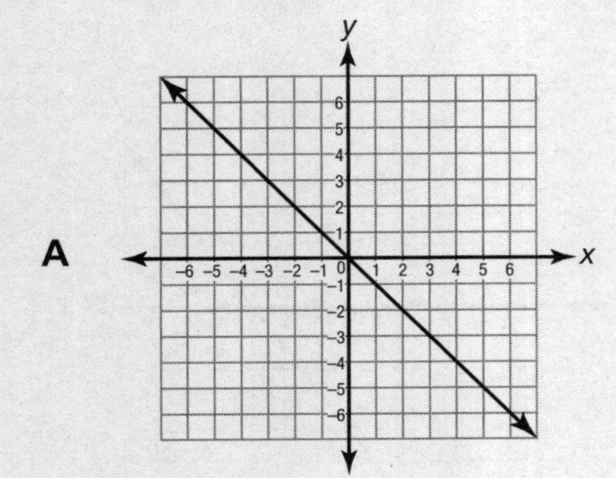

B
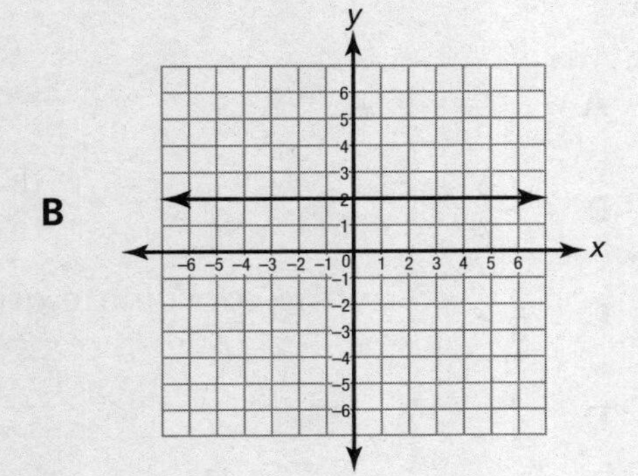

C
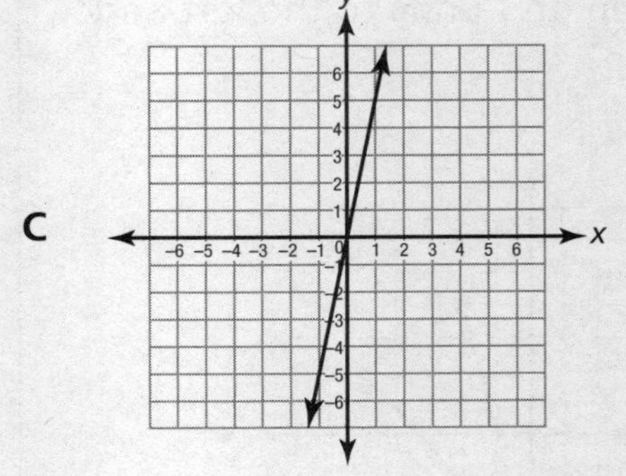

D
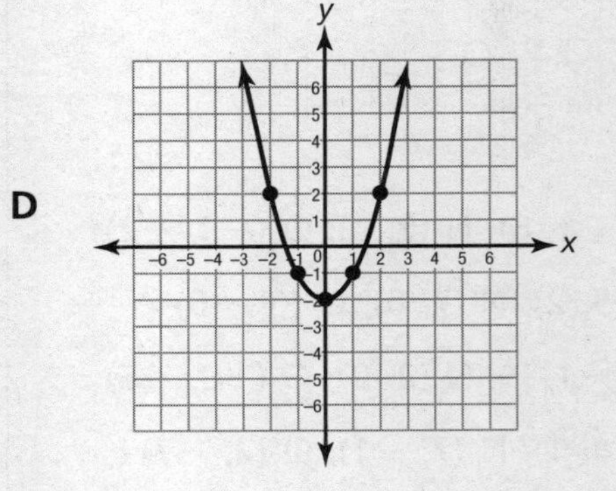

17 Which of the following is the graph of $y = -\frac{3}{2}x + 4$?

A
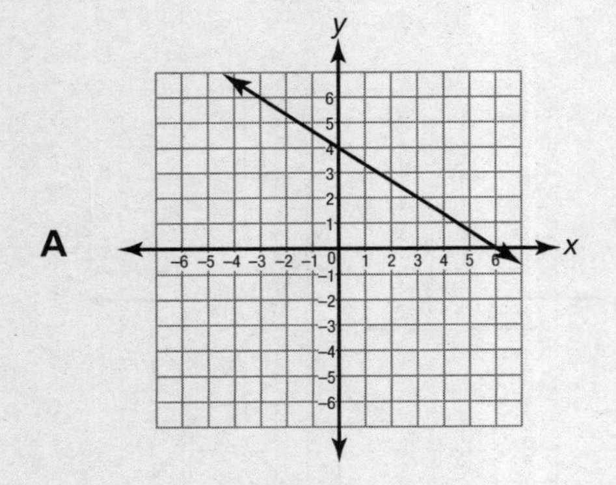

B
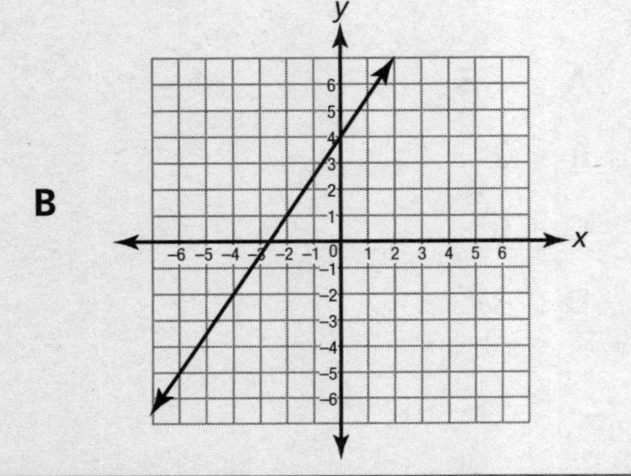

C
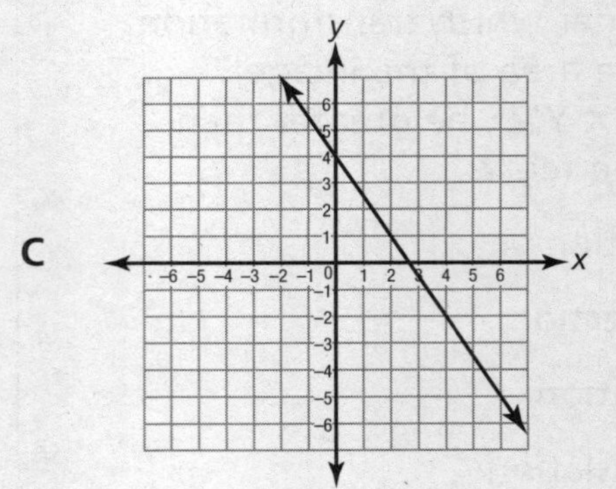

D
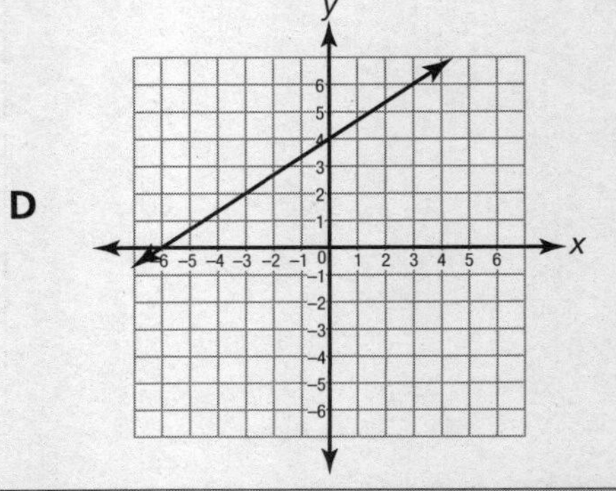

18 In the diagram below, ∠EOD and ∠EOF are complementary. What is the measure of ∠EOF?

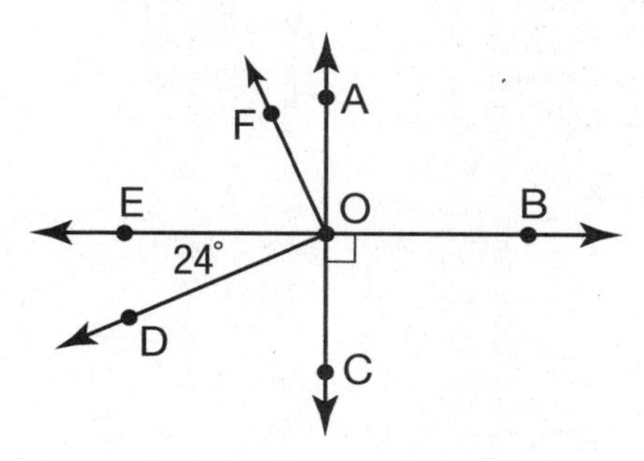

Show your work.

Answer ______________

19 Point A is rotated 180° about the origin. Plot image point A′ on the coordinate grid below and give its coordinates.

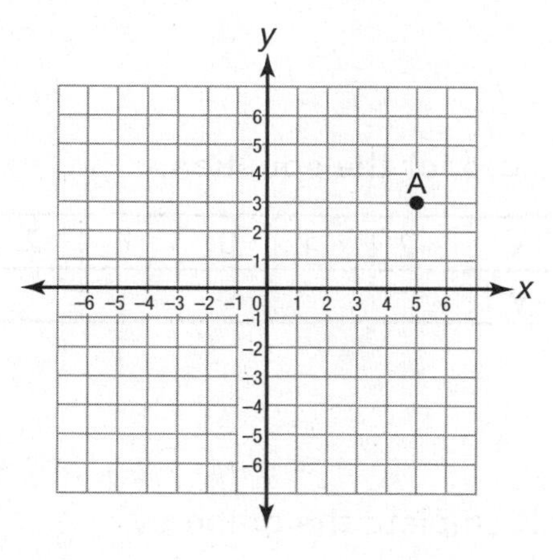

Answer (______________ , ______________)

20 Construct the perpendicular bisector of segment JK below.

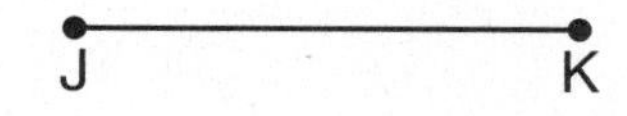

21 In the diagram below, lines AB and CD are parallel. Line EF is a transversal.

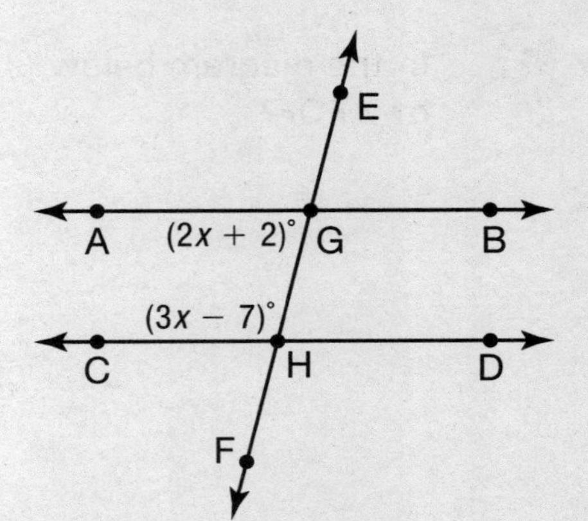

Part A

Find the measure of $\angle CHG$.

Answer __________

Part B

Explain how you determined your answer to *Part A*.

____

____

22 Consider the equation $y = -2x + 2$.

Part A

Complete this table of values for the equation.

x	-2	-1	0	1	2
y					

Part B

Graph the ordered pairs. Complete the graph by drawing a line through the points.

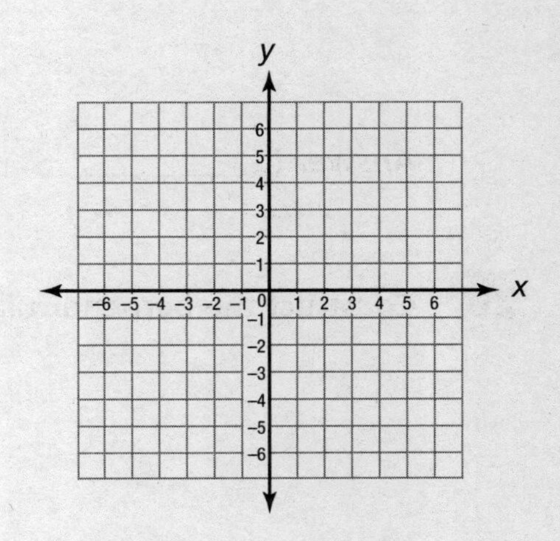

STRAND

4 Measurement

* Grade 7 May–June Indicators

Lesson

29 Convert Customary and Metric Units

Getting the Idea

Two well-known measurement systems are the **customary system** and the **metric system**. The customary system is used in the United States, while the metric system is used in most of the rest of the world.

The table below shows some conversions of length for the two systems.

Customary Units of Length	Metric Units of Length
1 foot (ft) = 12 inches (in.) 1 yard (yd) = 3 feet (ft) 1 yard = 36 inches (in.) 1 mile = 1,760 yards (yd) 1 mile (mi) = 5,280 feet (ft)	1 centimeter (cm) = 10 millimeters (mm) 1 meter (m) = 100 centimeters (cm) 1 meter = 1,000 millimeters 1 kilometer (km) = 1,000 meters (m)

To convert a larger unit to a smaller unit, multiply.

To convert a smaller unit to a larger unit, divide.

EXAMPLE 1

The track at the high school is a $\frac{1}{4}$-mile oval. Nicole ran 7 laps around the track. How many yards did she run?

STRATEGY **Convert miles to yards, then multiply by the number of laps. Multiply to convert a larger unit to a smaller unit.**

STEP 1 Write the relationship between miles and yards.

1 mile = 1,760 yards

STEP 2 Multiply the number of miles by 1,760 to find the number of yards.

$\frac{1}{4} \times 1{,}760 = 440$ yards

STEP 3 Multiply by the number of laps.

$440 \times 7 = 3{,}080$ yards

SOLUTION **Nicole ran 3,080 yards.**

When you want to know how heavy something is, you measure its weight. Weight is part of the customary system of measure. When you want to know how much matter an object has, you measure its mass. Mass is part of the metric system of measure. Unlike weight, which can vary according to location, mass never changes.

The table below shows some conversions of units of weight and mass.

Customary Units of Weight	Metric Units of Mass
1 pound (lb) = 16 ounces 1 ton (T) = 2,000 pounds (lb)	1 gram (g) = 1,000 milligrams (mg) 1 kilogram (kg) = 1,000 grams 1 metric ton (t) = 1,000 kilograms (kg)

A proportion can be useful for solving problems involving measurement conversion. For example, to find the number of pounds in 136 ounces, you could write this proportion:

$$\frac{16 \text{ ounces}}{1 \text{ pound}} = \frac{136 \text{ ounces}}{x \text{ pounds}}$$

Then use cross multiplication (the product of the means equals the product of the extremes) to write an equation and solve the equation. In the proportion above, the means are 1 and 136, and the extremes are 16 and x.

$1 \cdot 136 = 16 \cdot x$ $\quad 136 = 16x$ $\quad \frac{136}{16} = \frac{16x}{16}$ $\quad 8.5 = x$

There are 8.5 pounds in 136 ounces.

EXAMPLE 2

A hardcover book has a mass of 1.105 kilograms. A box of paper clips has a mass of 150 grams. How much more mass does the book have than the box of paper clips?

STRATEGY **Convert 1.105 kilograms to grams and subtract.**

STEP 1 Write and solve a proportion.

$$\frac{1{,}000 \text{ g}}{1 \text{ kg}} = \frac{x \text{ grams}}{1.105 \text{ kg}}$$

$1 \cdot x = 1{,}000 \cdot 1.105$

$x = 1{,}105$

The book has a mass of 1,105 g.

STEP 2 Subtract.

1,105 g − 150 g = 955 g

SOLUTION **The book has 955 grams more mass than the box of paper clips.**

Capacity measures the dry or liquid volume a container can hold. This table shows some capacity conversions in the customary and metric systems.

Customary Units of Capacity	Metric Units of Capacity
1 cup (c) = 8 fluid ounces (fl oz) 1 pint (pt) = 2 cups (c) 1 quart (qt) = 2 pints (pt) 1 gallon (gal) = 4 quarts (qt)	1 liter (L) = 1,000 milliliters (mL)

EXAMPLE 3

Jerry prepared 15.5 liters of sports drink for the football team to drink during the game. Rebecca prepared 18.75 liters. How many milliliters did they prepare in all?

STRATEGY **Convert liters to milliliters using multiplication or a proportion. Then add the totals.**

STEP 1 Convert 15.5 liters to milliliters.

Using multiplication:

$15.5 \times 1{,}000 = 15{,}500$ milliliters

So, 15.5 liters = 15,500 milliliters.

STEP 2 Convert 18.75 liters to milliliters.

Using a proportion:

$$\frac{1{,}000 \text{ mL}}{1 \text{ L}} = \frac{x \text{ mL}}{18.75 \text{ L}}$$

$1 \cdot x = 1{,}000 \cdot 18.75$

$x = 18{,}750$

So, 18.75 liters = 18,750 milliliters.

STEP 3 Add the milliliters.

$15{,}500 + 18{,}750 = 34{,}250$

SOLUTION **Jerry and Rebecca prepared 34,250 milliliters of sports drink.**

Just as you can convert between units of length, you can also convert between units of area (square units) and between units of volume (cubic units). You can use abbreviations for units of area and volume. For example:

square inches → sq in.

cubic centimeters → cu cm

Customary Units of Area	Metric Units of Area
144 sq in. = 1 sq ft 9 sq ft = 1 sq yd	100 sq mm = 1 sq cm 10,000 sq cm = 1 sq m 1,000,000 sq m = 1 sq km

Customary Units of Volume	Metric Units of Volume
1,728 cu in. = 1 cu ft 27 cu ft = 1 cu yd	1,000 cu mm = 1 cu cm 1,000,000 cu cm = 1 cu m

You can also use exponents to express units of area and volume. For example:

square inches → in.^2

cubic centimeters → cm^3

EXAMPLE 4

A hallway carpet covers 2,160 square inches of floor space. How many square feet of the hallway floor does the carpet cover?

STRATEGY **Convert square inches to square feet.**

Use a proportion.

$$\frac{1 \text{ ft}^2}{144 \text{ in.}^2} = \frac{x \text{ ft}^2}{2{,}160 \text{ in.}^2}$$

$$144 \cdot x = 1 \cdot 2{,}160$$

$$144x = 2{,}160$$

$$\frac{144x}{144} = \frac{2{,}160}{144}$$

$$x = 15$$

SOLUTION **The carpet covers 15 square feet of the hallway floor.**

Note: In Example 4 you could also have used division to convert from the smaller unit (in.^2) to the larger unit (ft^2).

COACHED EXAMPLE

How many fluid ounces are in 3 pints?

THINKING IT THROUGH

One pint = _____ cups, so 3 pints = ______ cups.

How many ounces are in 1 cup? ______

Multiply the number of ounces in 1 cup by the number of cups in 3 pints.

____

There are ________ fluid ounces in 3 pints.

Lesson Practice

Choose the correct answer.

1. The ceiling in Mrs. Abram's class is 12 feet 4 inches high. How many inches high is the ceiling?

A. 124 inches
B. 136 inches
C. 140 inches
D. 148 inches

2. A bottle can hold 175 mL of water. How many liters of water can it hold?

A. 0.00175 L
B. 0.175 L
C. 1,750 L
D. 17,500 L

3. The mass of a small dog is 6.5 kilograms. The mass of its toy is 125 grams. How much greater is the mass of the dog than the mass of its toy?

A. 6.0375 kg
B. 6.375 kg
C. 6.625 kg
D. 118.5 kg

4. A soup recipe calls for 3 quarts of water. Troy's only measuring cup can hold 1 cup of water. How many times must he fill the measuring cup to obtain enough water for the recipe?

A. 6
B. 12
C. 18
D. 24

5. Kim weighed 102 ounces when she was born. What was Kim's weight in pounds and ounces?

A. 6 pounds 6 ounces
B. 7 pounds
C. 8 pounds 6 ounces
D. 10 pounds 2 ounces

6. A square bathroom tile has an area of 16 square centimeters. What is the tile's area in square millimeters?

A. 0.16 sq mm
B. 160 sq mm
C. 1,600 sq mm
D. 16,000 sq mm

7. A box has a volume of 75 cubic centimeters. What is the volume of the box in cubic millimeters?

A. 750 mm^3
B. 7,500 mm^3
C. 75,000 mm^3
D. 750,000 mm^3

8. Elm Street is $\frac{1}{2}$ mile long. Spruce Street is 2,500 feet long. How many feet longer is Elm Street?

Answer ____________

EXTENDED-RESPONSE QUESTION

9. A fruit punch recipe that makes two gallons of punch calls for 1 gallon of orange juice, 2 quarts of apple juice, and 3 pints of grape juice. The rest is pineapple juice.

Part A

How many pints of pineapple juice are needed?

Answer ________

Part B

Explain how you determined your answer.

____

____

____

Lesson

30 Convert Temperatures

Getting the Idea

The most commonly used temperature scales are **Fahrenheit** (F) and **Celsius** (C). The temperature 75°F is read "seventy-five degrees Fahrenheit." The temperature 20°C is read "twenty degrees Celsius."

To convert between these two temperature scales, use the formulas below.

$F = \frac{9}{5}C + 32$ $\qquad$ $C = \frac{5}{9}(F - 32)$

EXAMPLE 1

The extreme high temperature recorded for Albany, New York, is 100°F. What is this extreme high temperature in degrees Celsius?

STRATEGY **Use the formula to convert Fahrenheit to Celsius.**

STEP 1 Write the formula.

$C = \frac{5}{9}(F - 32)$

STEP 2 Replace *F* with 100 and simplify.

$C = \frac{5}{9}(100 - 32)$

$= \frac{5}{9}(68)$

$= 37\frac{7}{9} \approx 37.8$

SOLUTION **The extreme high temperature recorded in Albany is about 37.8°C.**

EXAMPLE 2

On one day in January, the low temperature in Rochester, New York, was $-15°C$. What is this temperature in degrees Fahrenheit?

STRATEGY **Use the formula to convert Celsius to Fahrenheit.**

STEP 1 Write the formula.

$$F = \frac{9}{5}C + 32$$

STEP 2 Replace C with -15 and simplify.

$$\begin{aligned} F &= \frac{9}{5}C + 32 \\ &= \frac{9}{5}(-15) + 32 \\ &= -27 + 32 \\ &= 5 \end{aligned}$$

SOLUTION **The low temperature that day in Rochester was 5°F.**

EXAMPLE 3

Justin is planning a vacation in Daytona Beach, Florida. He can visit in January, May, or September, but wants to visit when the monthly average high temperature is less than 30°C.

Daytona Beach Monthly Average High Temperatures

Month	Average High Temperature
January	70°F
May	85°F
September	88°F

During which month or months should Justin plan to go to Daytona Beach?

STRATEGY **Use the formula to convert Fahrenheit to Celsius and then compare.**

STEP 1 Convert January's average high temperature to Celsius.

$$\begin{aligned} C &= \frac{5}{9}(F - 32) \\ &= \frac{5}{9}(70 - 32) \\ &= \frac{5}{9}(38) \\ C &\approx 21.1°C \end{aligned}$$

STEP 2 Convert May's average high temperature to Celsius.

$$C = \frac{5}{9}(F - 32)$$
$$= \frac{5}{9}(85 - 32)$$
$$= \frac{5}{9}(53)$$
$$C \approx 29.4°\text{C}$$

STEP 3 Convert September's average high temperature to Celsius.

$$C = \frac{5}{9}(F - 32)$$
$$= \frac{5}{9}(88 - 32)$$
$$= \frac{5}{9}(56)$$
$$C \approx 31.1°\text{C}$$

STEP 4 Compare each result to 30°C.

January: $21.1°\text{C} < 30°\text{C}$

May: $29.4°\text{C} < 30°\text{C}$

September: $31.1°\text{C} > 30°\text{C}$

SOLUTION **Justin should plan to take a vacation in Daytona Beach during January or May.**

COACHED EXAMPLE

The average high temperature in July in New York City is 25°C. What is the average high temperature in July in New York City in degrees Fahrenheit?

THINKING IT THROUGH

Use the formula $F =$ ______________.

Replace C with _________ and simplify.

Compute to find F.

$F =$ ____

The average high temperature in July in New York City is _____°F.

Lesson Practice

Choose the correct answer.

1. What is 40°C in degrees Fahrenheit?

   A. 40°F
   B. 50°F
   C. 72°F
   D. 104°F

2. What is 50°F in degrees Celsius?

   A. 10°C
   B. 18°C
   C. 50°C
   D. 122°C

3. Joe preheated his oven to 375°F. To the nearest ten, what was the oven temperature in degrees Celsius?

   A. 130°C
   B. 150°C
   C. 170°C
   D. 190°C

4. Johanna spent the day at the beach, where the temperature reached 32°C. What is 32°C in degrees Fahrenheit?

   A. 89.6°F
   B. 57.6°F
   C. 32°F
   D. 0°F

5. Ricardo went skiing, where the temperature reached a low of 8°F. What is 8°F in degrees Celsius?

   A. about 13.3°C
   B. about −8.7°C
   C. about −13.3°C
   D. about −24°C

6. Which pair has a temperature less than 90°F and a temperature greater than 90°F?

   A. 30°C, 31°C
   B. 31°C, 32°C
   C. 32°C, 33°C
   D. 33°C, 34°C

7. Madeline wants to visit a Florida city in January, where the average low temperature is between 9°C and 12°C. Which two cities in the table should she visit?

January Average Low Temperature

City	Temperature
Key West	65°F
Naples	53°F
Orlando	50°F
Saint Augustine	46°F

   A. Key West and Naples
   B. Naples and Orlando
   C. Orlando and Saint Augustine
   D. Key West and Saint Augustine

8. The record high temperature in the state of New York is 108°F. To the nearest whole degree, what is this record temperature in degrees Celsius?

Answer __________

EXTENDED-RESPONSE QUESTION

9. The temperature in Andy's town is 75°F. The same day, the temperature in the Spanish town where his friend lives is 25°C.

Part A

Which temperature is greater, 75°F or 25°C?

Answer __________

Part B

Explain how you determined the greater temperature.

____

____

Lesson

31 Convert Currency

7.M.7

Getting the Idea

By using an exchange rate table, you can convert money between different currencies.

EXAMPLE 1

Ella went on a trip to England to visit relatives. She exchanged \$100 US for British pounds. How many British pounds did she receive?

U.S. Dollar	British Pound
1	0.617222

STRATEGY **Use the exchange rate table.**

STEP 1 Find the exchange rate between \$1 US and the British pound.

The symbol for the British pound is £.

\$1 US = £0.617222

STEP 2 Write a proportion comparing US dollars to British pounds.

Let p represent the number of British pounds.

$$\frac{\$1}{£0.617222} = \frac{\$100}{p}$$

STEP 3 Solve the proportion for p.

$1 \times p = 100 \times 0.617222$

$p = 61.7222$

STEP 4 Round to the nearest hundredth.

61.7222 rounds to 61.72.

SOLUTION **Ella received £61.72.**

You can also use a calculator to convert money.

EXAMPLE 2

Ricky went on a trip to Canada with his family last summer. His father exchanged $200 US for Canadian dollars. How many Canadian dollars did his father receive?

U.S. Dollar	Canadian Dollar
1	1.16625

STRATEGY **Use the exchange rate table.**

STEP 1 Find the exchange rate between $1 US and the Canadian dollar.

$1 US = $1.16625 Canadian

STEP 2 Use a calculator to multiply $200 by 1.16625.

$200 × 1.16625 = 233.25 Canadian dollars

SOLUTION **Ricky's father received $233.25 in Canadian dollars.**

EXAMPLE 3

Doriano's grandparents live in Italy. They came to visit him in New York last year. When they arrived, they exchanged 500 euros for U.S. dollars. How many U.S. dollars did they receive?

U.S. Dollar	European Euro
1.3901	1

STRATEGY **Use the exchange rate table.**

STEP 1 Find the exchange rate between the European euro and the U.S. dollar. The symbol for the European euro is €.

€1 = $1.3901

STEP 2 Use a calculator to multiply 500 euros by 1.3901.

500 × 1.3901 = $695.05

SOLUTION **Doriano's grandparents received 695.05 in U.S. dollars.**

COACHED EXAMPLE

Luke traveled in Europe and England last summer. When he arrived in England, he converted 200 euros into British pounds using the conversion table below.

British Pound	European Euro
0.847999	1

How many British pounds did Luke receive?

THINKING IT THROUGH

The exchange rate between euros and British pounds is €1 = £ ______________.

Write a proportion comparing euros to British pounds. Let p represent the number of British pounds.

______________ = ________________

Cross multiply and find the value of p.

__________ $\times p =$ __________ $\times$ __________

$p =$ __________

The number of British pounds rounded to the nearest hundredth is ____________.

Luke received £____________.

Lesson Practice

Choose the correct answer.

Use the exchange rate table below for questions 1–3.

Currency	Exchange Rate
U.S. Dollar	1
Chinese Yuan	6.8327
European Euro	0.719373
Australian Dollar	1.29055

1. Sophie exchanged $250 US for European euros when she arrived in France. How many euros did she get?

A. €1,798.43
B. €347.52
C. €179.84
D. €34.75

2. Jeremy will be traveling in Australia for one month. He exchanged $500 US for Australian dollars. How much money in Australian dollars did he get?

A. 387.43 Australian dollars
B. 645.28 Australian dollars
C. 3,874.32 Australian dollars
D. 6,452.75 Australian dollars

3. Ava exchanged $150 US for Chinese yuan when she arrived in China. How many yuan did she get?

A. 21.95 yuan
B. 124.91 yuan
C. 1,024.91 yuan
D. 10,249.05 yuan

4. Francesca traveled to Argentina to visit her cousins. She spent 25 Argentine pesos on some postcards to send to her friends in New York. How much in US dollars did the postcards cost if the exchange rate was 1 Argentine peso = $0.260247 U.S. dollars?

A. $6.51
B. $9.61
C. $65.06
D. $96.06

5. The price of admission at the Russian Museum in St. Petersburg, Russia, for a student ticket is 300 rubles. About how much in U.S. dollars is a student ticket if 1 Russian ruble = $0.030453 U.S. dollars?

A. $1
B. $9
C. $10
D. $90

6. Bruno lives in Italy. He went to England on vacation and exchanged 300 European euros for British pounds. If the exchange rate was €1 = £0.867799, how many British pounds did he get, to the nearest whole number?

Answer ____________________

7. Leah lives in Canada. She went to Mexico on vacation and exchanged 50 Canadian dollars for Mexican pesos. If the exchange rate was 1 Canadian dollar = 11.7973 pesos, how many pesos did she get, to the nearest whole number?

Answer ____________________

Lesson

32 Calculate and Compare Unit Prices

7.M.5, 7.M.6

Getting the Idea

The **unit price** of an item is its price per ounce, per quart, or per any unit of measure. You can use a proportion to find the unit price of an item.

$$\frac{\text{unit price}}{1 \text{ unit}} = \frac{\text{price of item}}{\text{number of units in item}}$$

The unit price of an item is sometimes expressed to the tenth of a cent.

EXAMPLE 1

Jemma bought a 5-pound bag of potatoes for \$2.29. What is the unit price for one pound of potatoes?

STRATEGY **Write and solve a proportion.**

STEP 1 Write a proportion.

The price of the item is \$2.29 and the number of units in the item is 5.

$$\frac{\text{unit price}}{1 \text{ unit}} = \frac{\$2.29}{5}$$

Let x stand for the unknown unit price.

$$\frac{x}{1} = \frac{\$2.29}{5}$$

STEP 2 Cross-multiply and solve for x.

$$\frac{x}{1} = \frac{\$2.29}{5}$$

$$5 \times x = \$2.29 \times 1$$

$$5x = \$2.29$$

$$5x \div 5 = \$2.29 \div 5$$

$$x = \$0.458, \text{ or } 45.8¢$$

SOLUTION **The unit price for one pound of potatoes is \$0.458, or 45.8¢.**

You can use unit pricing to compare the prices of two different sizes of the same item. The better buy is the item that has the lower unit price. You may have to divide the dollar amount to the ten-thousandths place, then change the dollar amount to cents, and round to the nearest tenth of a cent.

EXAMPLE 2

A 64-ounce bottle of juice costs $2.49 and a 40-ounce bottle of the same juice costs $1.69. Which is the better buy?

STRATEGY **Find the unit price of each bottle of juice and compare the unit prices.**

STEP 1 Find the unit price for the 64-ounce bottle.

$$\frac{\text{unit price}}{1\text{ unit}} = \frac{\$2.49}{64}$$

$$\frac{x}{1} = \frac{\$2.49}{64}$$

$$64x = \$2.49$$

Divide to the ten-thousandths place.

$x = \$2.49 \div 64 = \$0.0389$, or 3.89¢

Round to the nearest tenth of a cent.

3.89¢ rounds to 3.9¢.

STEP 2 Find the unit price for the 40-ounce bottle.

$$\frac{\text{unit price}}{1\text{ unit}} = \frac{\$1.69}{40}$$

$$\frac{x}{1} = \frac{\$1.69}{40}$$

$$40x = \$1.69$$

Divide to the ten-thousandths place.

$x = \$1.69 \div 40 = \$0.0422$, or 4.22¢

Round to the nearest tenth of a cent.

4.22¢ rounds to 4.2¢.

STEP 3 Compare the unit prices.

3.9¢ $<$ 4.2¢

SOLUTION **Since the unit price for the 64-ounce bottle is less than the unit price for the 40-ounce bottle, the 64-ounce bottle is the better buy.**

COACHED EXAMPLE

Randy can buy an 8.2-ounce tube of toothpaste for \$2.99 or a 6.4-ounce tube of the same toothpaste for \$2.39. Which is the better buy?

THINKING IT THROUGH

Find the unit price for the 8.2-ounce tube of toothpaste.

Write a proportion that can be used to find the unit price. ____________________

Cross-multiply. ___________ = ___________

Divide to the ten-thousandths place.

$x =$ \$___________ or ___________¢

Round to the nearest tenth of a cent.

___________¢ rounds to ___________¢.

The unit price for the 8.2-ounce tube of toothpaste is ___________.

Find the unit price for the 6.4-ounce tube of toothpaste.

Write a proportion that can be used to find the unit price. ____________________

Cross-multiply. ___________ = ___________

Divide to the ten-thousandths place.

$x =$ \$___________ or ___________¢

Round to the nearest tenth of a cent.

___________¢ rounds to ___________¢.

The unit price for the 6.4-ounce tube of toothpaste is ___________.

___________¢ < ___________¢, so the ________-ounce tube of toothpaste is the better buy.

Lesson Practice

Choose the correct answer.

1. Camille bought 3 pounds of nuts for $10.35. What was the unit price per pound?

A. $3.45
B. $4.65
C. $6.75
D. $7.35

2. Luis bought a 6-ounce container of yogurt for $0.75. What was the unit price per ounce?

A. $1.25
B. 15¢
C. 12.5¢
D. 8¢

3. Which of these packages of Aunt Betty's Rice is the best buy?

A. a 12-ounce box for $1.59
B. a 16-ounce box for $2.09
C. a 32-ounce box for $4.29
D. an 80-ounce box for $10.49

4. Sam bought a 14.5-ounce can of tomatoes for $1.39. What was the unit price per ounce, to the nearest tenth of a cent?

A. 9.0¢
B. 9.2¢
C. 9.6¢
D. 9.9¢

5. Which of the following shows the lowest unit price per orange?

A. 3 oranges for $1.02
B. 4 oranges for $1.52
C. 5 oranges for $1.75
D. 6 oranges for $2.46

6. Which of the following shows the lowest unit price per pound of bananas?

A. 4 pounds for $2.29
B. 5 pounds for $2.99
C. 2 pounds for $1.19
D. 3 pounds for $1.69

7. Mykala bought a 12-ounce box of pasta for $0.89. What was the unit price per ounce, to the nearest tenth of a cent?

A. 7.0¢
B. 7.4¢
C. 7.5¢
D. 8.0¢

8. Which of these boxes of Cracklin' Corn Cereal is the best buy?

A. a 12.2-ounce box for $2.59
B. a 16-ounce box for $3.29
C. a 19.8-ounce box for $4.29
D. a 25.4-ounce box for $4.99

9. Which is the better buy: a 7-ounce bottle of shampoo for $2.59, or a 12-ounce bottle of shampoo for $4.29?

Answer ____________________

10. Which is the better buy: a 32-load container of detergent for $4.89, or a 50-load container of detergent for $7.99?

Answer ____________________

EXTENDED-RESPONSE QUESTION

11. A 12-ounce box of Crispy Crackers costs $1.89, while a 16-ounce box of Crispy Crackers costs $2.29.

Part A Which is the better buy—the 12-ounce box or the 16-ounce box?

____

Part B Explain how you figured out which was the better buy. Be sure to mention the unit price for each box in your explanation.

____

____

____

Lesson

33 Map Scale

7.M.1

Getting the Idea

Some objects are too large or too small to be drawn to their actual sizes. In that case, you can create a **scale drawing**. Scale drawings, such as maps, use a **scale factor** that gives the ratio of the size of the drawing to the size of the actual object it represents. On a map, the scale provides the scale factor.

EXAMPLE 1

Desiree is using this map to find her way around a carnival. If she walks directly from the ticket booth to the dunk tank, how far will she walk?

Carnival Map

Ticket Booth
Face Painting
Dunk Tank
Cake Walk
Refreshments

Scale: 1 centimeter = 90 meters

STRATEGY **Set up and solve a proportion.**

STEP 1 Write the scale factor as a ratio.

The scale is 1 centimeter = 90 meters.

This can be written as $\frac{1 \text{ cm}}{90 \text{ m}}$.

STEP 2 Use a ruler to measure the straight-line distance between the ticket booth and the dunk tank on the map.

The distance is 4.5 cm.

STEP 3 Write a ratio to represent the scale drawing length and the unknown actual length.

$$\frac{\text{drawing length}}{\text{actual length}} = \frac{4.5 \text{ cm}}{x \text{ m}}$$

STEP 4 Set up and solve a proportion to find x.

Each ratio compares centimeters to meters. Set the two ratios equal and solve.

$$\frac{1 \text{ cm}}{90 \text{ m}} = \frac{4.5 \text{ cm}}{x \text{ m}}, \text{ or } \frac{1}{90} = \frac{4.5}{x}$$

$$1 \times x = 90 \times 4.5$$

$$x = 405$$

SOLUTION **If Desiree walks from the ticket booth to the dunk tank, she will walk 405 meters.**

EXAMPLE 2

The distance between the gym and the main entrance of a school is 7 inches on a map. The scale of the map is 2 inches = 75 feet. What is the actual distance between the main entrance and the gym?

STRATEGY **Write and solve a proportion.**

STEP 1 Write the scale factor as a ratio.

The scale of the map is 2 inches = 75 feet.

This can be written as $\frac{2 \text{ in.}}{75 \text{ ft}}$.

STEP 2 Write a ratio to represent the scale drawing length and the unknown actual length.

Let x represent the unknown number of feet.

$$\frac{\text{drawing length}}{\text{actual length}} = \frac{7 \text{ in.}}{x \text{ ft}}$$

STEP 3 Set up and solve a proportion to find x.

Each ratio compares inches to feet. Set the two ratios equal and solve.

$$\frac{2 \text{ in.}}{75 \text{ ft}} = \frac{7 \text{ in.}}{x \text{ ft}}, \text{ or } \frac{2}{75} = \frac{7}{x}$$

$$2 \times x = 75 \times 7$$

$$2x = 525$$

$$x = 262.5$$

SOLUTION **The actual distance between the main entrance and the gym is 262.5 feet.**

COACHED EXAMPLE

Sergio attends a community college. Use the map of the community college below to find the actual distance from the education building to the food court.

Map of Community College

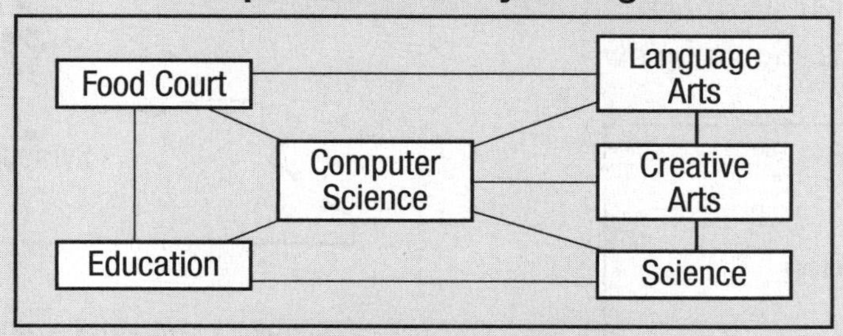

Scale: 0.5 cm = 20 m

THINKING IT THROUGH

Write a ratio to represent the scale. __________

Measure the length of the line segment connecting the food court and the education building on the map with a centimeter ruler.

The distance is ________ centimeters.

Write a second ratio that compares the scale drawing length you just found to *d*, the unknown actual distance between the education building and the food court:

$\frac{\text{drawing length}}{\text{actual length}} =$ __________

Set up a proportion and use it to solve for *d*.

The actual distance from the education building to the food court is ______ meters.

Lesson Practice

Choose the correct answer.

1. Use your inch ruler to help you answer this question.

   A scale drawing of Mr. Farmer's rectangular garden is shown below.

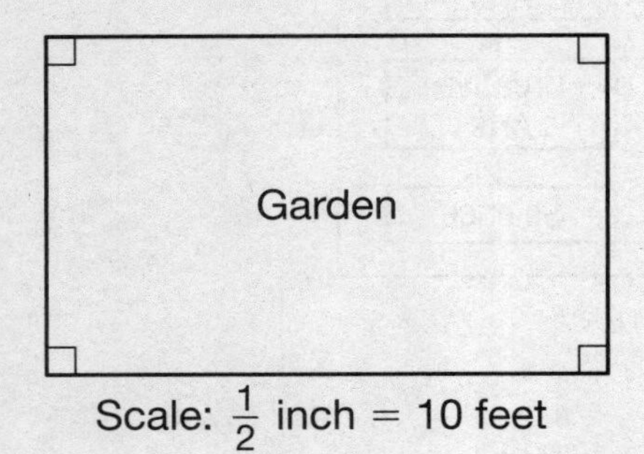

   What are the actual dimensions of his garden?

   A. 20 feet by 12 feet
   B. 20 feet by 16 feet
   C. 40 feet by 20 feet
   D. 40 feet by 25 feet

2. A map has a scale of 2 inches = 15 miles. If two towns are 7 inches apart on the map, what is the actual distance between the towns?

   A. 52.5 miles
   B. 55 miles
   C. 57.5 miles
   D. 105 miles

Use the map of New York below and your inch ruler for questions 3 and 4.

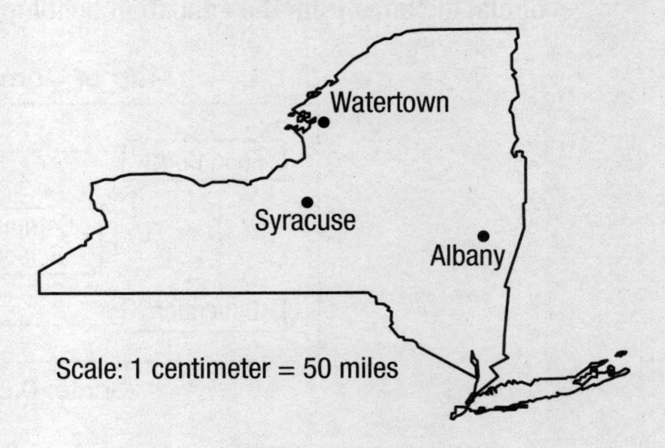

3. What is the actual distance between Watertown and Syracuse?

   A. 25 miles
   B. 50 miles
   C. 100 miles
   D. 125 miles

4. What is the actual distance between Watertown and Albany?

   A. 100 miles
   B. 125 miles
   C. 175 miles
   D. 225 miles

5. On a map, 1 centimeter represents 20 kilometers. Two rivers on the map are 8.1 centimeters apart at a certain location. What is the actual distance between the two rivers at that location?

A. 1,620 kilometers

B. 178.2 kilometers

C. 168 kilometers

D. 162 kilometers

6. On a map of New York state, the distance between Oneonta and Jamestown is 5.6 cm. The scale is 1 cm = 50 mi. What is the actual distance between Oneonta and Jamestown?

A. 256 miles

B. 280 miles

C. 300 miles

D. 560 miles

Use the following information for questions 7 and 8.

The gym teachers at Franklin Middle School organized Field Day. The map below shows the locations of different Field Day activities. Use your inch ruler to help you answer the questions.

Field Day Activities

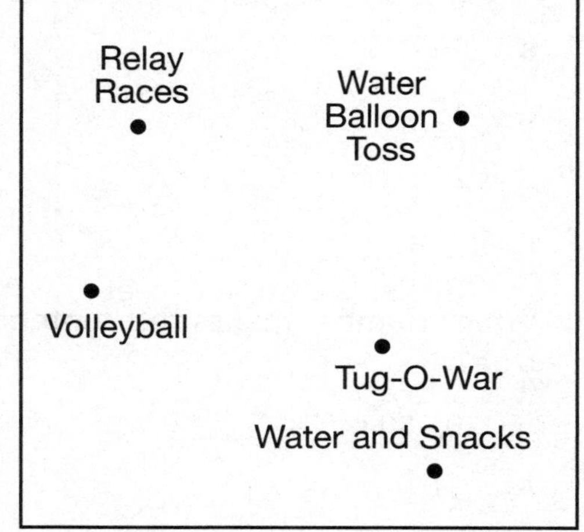

Scale: $\frac{1}{4}$ inch = 15 yards

7. How many yards is it from the tug-o-war to the relay races?

Answer ____________________

8. After playing volleyball, the students walked to Water and Snacks. How many yards did they walk?

Answer ____________________

4 Review

1 Which number makes this sentence true?

10 qt = ____ pt

A 5

B 10

C 15

D 20

2 Which number makes this sentence true?

10,000 mg = ______ kg

A 0.01

B 0.1

C 1

D 10

3 What is 41°F in degrees Celsius?

A 0°C

B 5°C

C 10°C

D 15°C

4 What is 20°C in degrees Fahrenheit?

A 4°F

B 52°F

C 68°F

D 77°F

5 A rectangular photograph is 15 cm long and 10 cm wide. What is the area of the photo in square millimeters?

A 150 mm^2

B 250 mm^2

C 1,500 mm^2

D 15,000 mm^2

6 A jug holds 2.36 L of lemonade. How many milliliters of lemonade does the jug hold?

A 0.0236 mL

B 236 mL

C 2,360 mL

D 236,000 mL

7 A storage room has a volume of 12 cubic yards. What is the volume of the storage room in cubic feet?

A 36 ft^3

B 108 ft^3

C 144 ft^3

D 324 ft^3

8 A boat is 25 feet 6 inches long. What is the length of the boat in inches?

A 256 in.

B 300 in.

C 306 in.

D 360 in.

9 Two cities are shown on a map. The scale on the map is shown below.

SCALE
1 cm = 32 km

The actual distance between the two cities is 240 kilometers. How many centimeters are the two cities apart on the map?

A 7.5 centimeters

B 208 centimeters

C 272 centimeters

D 7,680 centimeters

10 A 15-ounce can of soup costs $1.95. What is the can of soup's price per ounce?

A $0.11

B $0.13

C $0.17

D $0.19

11 Mr. McGrath is buying a box of pasta for his family's dinner. There are four different box sizes of the same brand of pasta. Which is the **best** buy?

A a 12-ounce box for $1.59

B a 16-ounce box for $2.09

C a 32-ounce box for $4.29

D an 80-ounce box for $10.49

12 Bella went to visit her family in Seville, Spain. The exchange rate from U.S. dollars to euros was \$1 U.S. = €0.70 EUR.

Bella has \$550 to convert into euros. How many euros will she receive?

Show your work.

Answer € ______________

13 Pedro says water boils at 212°F. What is 212°F in degrees Celsius?

Show your work.

Answer __________

14 A spool contains 24 yards of wire. A crafts project requires pieces of wire that are 9 inches long.

Part A

How many 9-inch pieces of wire can be cut from the spool?

Answer __________

Part B

Explain how you determined your answer to *Part A*.

____

Glossary

adjacent angles two angles in a plane that share a common side and a common vertex, but have no interior points in common (Lesson 18)

alternate exterior angles a pair of angles on the outer sides of two lines intersected by a transversal, but on opposite sides of the transversal; if the two lines are parallel, alternate exterior angles are congruent (Lesson 19)

alternate interior angles a pair of angles on the inner sides of two lines intersected by a transversal, but on opposite sides of the transversal; if the two lines are parallel, alternate interior angles are congruent (Lesson 19)

base a number that is raised to an exponent (Lesson 1)

binomial a polynomial with two unlike terms (Lesson 9)

Celsius a temperature scale based on 0° as the freezing point of water and 100° as the boiling point of water (Lesson 30)

coefficient a constant that multiplies a variable (Lesson 5)

commission earnings based on a percent of total sales (Lesson 3)

complementary angles two angles whose measures have a sum of 90° (Lesson 18)

constant a quantity that does not change its value in a given expression or equation (Lesson 5)

construction a precise way of drawing that allows only two tools: a straightedge and a compass (Lesson 22)

corresponding angles a pair of angles, one interior and one exterior, formed by two lines intersected by a transversal; corresponding angles are on the same side of the transversal and are congruent if the two lines are parallel (Lesson 19)

customary system the system of measurement used mainly in the United States to measure length, weight, and capacity (Lesson 29)

dilation a transformation that enlarges or reduces a figure by a scale factor to form a similar figure (Lesson 21)

discount a reduction made from the regular or list price of something (Lesson 3)

distributive property a property that states that the product of a number and the sum or difference of two numbers is the same as the sum or difference of their respective products (Lesson 11)

domain the set of all values for which a function is defined; the set of all possible input values (Lesson 17)

estimate an approximation (Lesson 4)

exponent a number that tells how many times the base is used as a factor; in an expression of the form b^a, a is the exponent, b is the base, and b^a is a power of b (Lesson 1)

exponential form form of a number written using exponents (Lesson 1)

expression a mathematical representation that may contain numbers, variables, and operation symbols; an expression does not include an equality or inequality symbol (Lesson 5)

factor to express a quantity as a product of two or more other quantities (Lesson 12)

Fahrenheit a temperature scale based on 32° as the freezing point of water and 212° as the boiling point of water (Lesson 30)

FOIL method an algorithm for multiplying binomials; FOIL stands for First terms, Outside terms, Inside terms, and Last terms (Lesson 11)

function a relation in which each input value (x-value) corresponds to only one output value (y-value) (Lesson 17)

graph a graphic representation used to show a relationship between sets of data (Lesson 8)

image a figure created when another figure, the pre-image, undergoes a transformation (Lesson 20)

inequality a mathematical statement containing one of the symbols $>$, $<$, $\geq$, $\leq$, or $\neq$ to indicate the relationship between two quantities (Lesson 6)

interest rate a percent charged on money borrowed or earned on money invested (Lesson 3)

like terms terms that have the same variables raised to the same powers (Lesson 9)

linear equation an equation in which none of the terms are raised to a power greater than 1; the graph of a linear equation is a straight line (Lesson 27)

metric system a system of measurement based on the decimal system; the standard unit of length is a meter, of capacity is a liter, and of mass is a gram (Lesson 29)

monomial a polynomial with one term; it is a number, a variable, or the product of a number and one or more variables (Lesson 9)

nonlinear equation an equation whose graph is not a straight line (Lesson 27)

ordered pair a set of two numbers, usually represented by x and y (Lesson 8)

parabola the graph of a quadratic function (Lesson 28)

parallel lines lines in the same plane that remain equidistant and do not intersect (Lesson 19)

percent a ratio that compares a number to 100 (Lesson 2)

percent of decrease a decrease expressed as a percent of the original quantity (Lesson 3)

percent of increase an increase expressed as a percent of the original quantity (Lesson 3)

polynomial a monomial or the sum/difference of monomials whose exponents are positive (Lesson 9)

principal the original amount of money either invested or borrowed; excludes any accrued interest (Lesson 3)

quadratic a polynomial with a variable raised to the second power (Lesson 28)

quadratic equation a polynomial equation where no power is greater than 2; its general form is $ax^2 + bx + c = 0$, where x is the variable and a, b, and c are constants (Lesson 28)

quadratic function a function of the form $y = ax^2 + bx + c$, where x is the variable and a, b, and c are constants (Lesson 28)

range the set of output values (y-values) of a function (Lesson 17)

rate of change the amount that a function's output increases or decreases for each change in the input (Lesson 23)

reflection a transformation in which every point of a figure is flipped over a line to create a mirror image (Lesson 20)

relation a set of ordered pairs (Lesson 16)

rotation a transformation in which every point of a figure is turned about a fixed point; also called a turn (Lesson 20)

sale price the price of a product after the discount has been subtracted from the original price (Lesson 3)

sales tax an amount of money charged in addition to the cost of an item or service (Lesson 3)

scale drawing a representation of a place or object that has an exact ratio in size to that place or object (Lesson 33)

scale factor a ratio of the lengths of the sides of a figure and its image after a dilation (Lesson 21); the ratio of a model to the object it represents (Lesson 33)

simple interest the amount of money obtained by multiplying the principal by the rate by the time; $I = prt$ (Lesson 3)

slope the measure of the steepness of a line; it is the ratio of the vertical change to the corresponding horizontal change (Lesson 23)

slope-intercept form the equation of a straight line in the form $y = mx + b$, where m is the slope and b is the y-coordinate of the point where the line crosses the y-axis (Lesson 25)

solution set the set of values that make an inequality true (Lesson 7)

standard form a number in which each digit is in a place value (Lesson 1)

supplementary angles two angles whose measures have a sum of 180° (Lesson 18)

system of linear equations a set of linear equations that may or may not share one or more common solutions (Lesson 26)

term an addend in an algebraic expression; it may be a number, a variable, or the product of a number and one or more variables (Lesson 9)

transformation the result of a change made to an object; a transformation may be a translation, a rotation, a reflection, or a dilation (Lesson 20)

translation a transformation in which every point of a figure moves the same distance in one or two directions in a plane; sometimes called a slide (Lesson 20)

transversal a line that intersects two or more other lines (Lesson 19)

trinomial a polynomial with exactly three unlike terms (Lesson 9)

unit price the price of one unit of a quantity (Lesson 32)

variable a symbol used to represent a number or group of numbers in an expression, equation, or inequality (Lesson 5)

vertical angles the nonadjacent angles formed by two intersecting lines; vertical angles are congruent (Lesson 18)

vertical line test a test to determine whether a relation is a function; a relation is a function if there are no vertical lines that intersect the graph of the function at more than one point (Lesson 17)

***x*-intercept** the point where a graph of an equation crosses the x-axis (Lesson 24)

***y*-intercept** the point where a graph of an equation crosses the y-axis (Lesson 24)

New York State Coach, Empire Edition, Mathematics, Grade 8

PRACTICE TEST 1

Name: ______________________________

TIPS FOR TAKING THE TEST

Here are some suggestions to help you do your best:

- Be sure to read carefully all the directions in the test book.
- Read each question carefully and think about the answer before writing your response.
- Be sure to show your work when asked. You may receive partial credit if you have shown your work.
- Use your calculator to help you solve the problems in Sessions 2 and 3.

This picture means that you will use your ruler.

Session 1

PRACTICE TEST 1

1 What is $\frac{5^5}{5^3}$ in exponential form?

A 5^{-2}

B 5^{-8}

C 5^2

D 5^8

2 There were 200 raffle tickets sold. Jewel bought one raffle ticket. What is her chance of winning expressed as a percent?

A 0.2%

B 0.5%

C 2%

D 5%

3 If a polygon has 8 sides, which expression can you use to find the sum of the measures of the interior angles of the polygon?

A $8(180 - 2)$

B $2(180 - 8)$

C $180(8 - 2)$

D $180(8 + 2)$

4 Alice's mother told her that she could invite no more than 20 people, p, to her birthday party. What inequality represents what Alice's mother told her?

A $p \leq 20$

B $p \geq 20$

C $p < 20$

D $p > 20$

Go On

5 What is the sum of the polynomials modeled below?

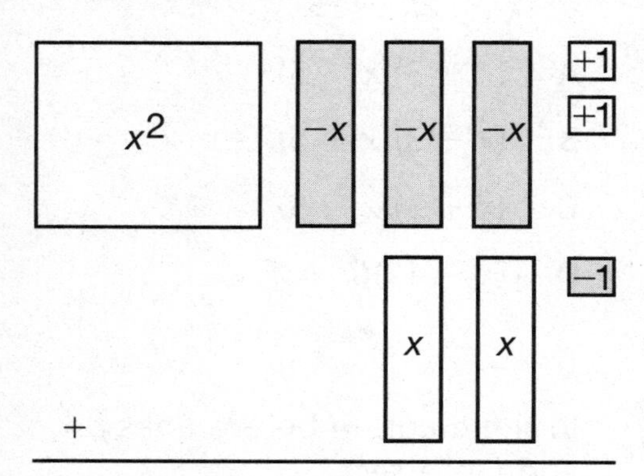

A $x^2 + 5x + 3$

B $x^2 - x + 1$

C $x^2 + x + 1$

D $x^2 - x - 1$

6 Simplify the expression below.

$7x(3x - 4)$

A $21x - 4$

B $21x - 28x$

C $21x^2 - 4$

D $21x^2 - 28x$

Go On

7 Which graph **best** represents walking from home to the store and then back home?

A

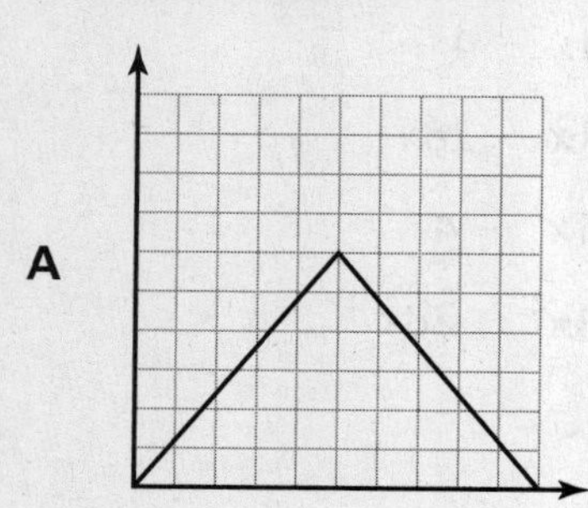

B

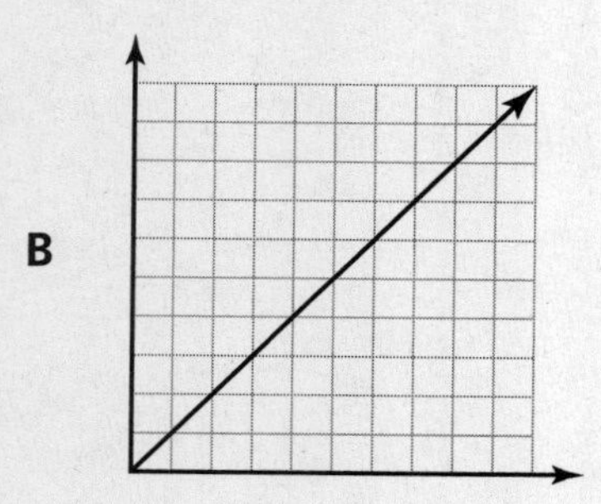

C

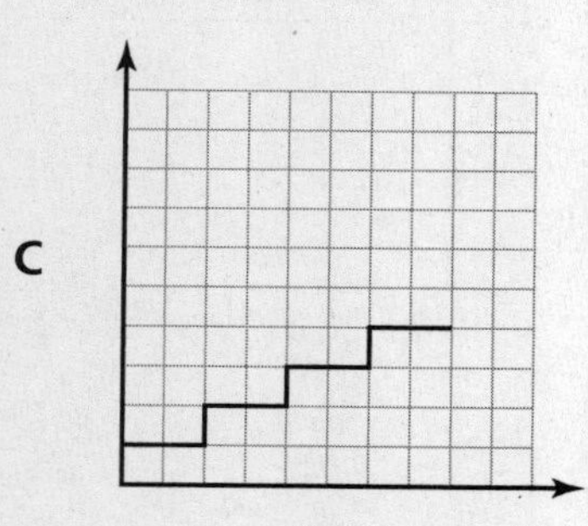

D

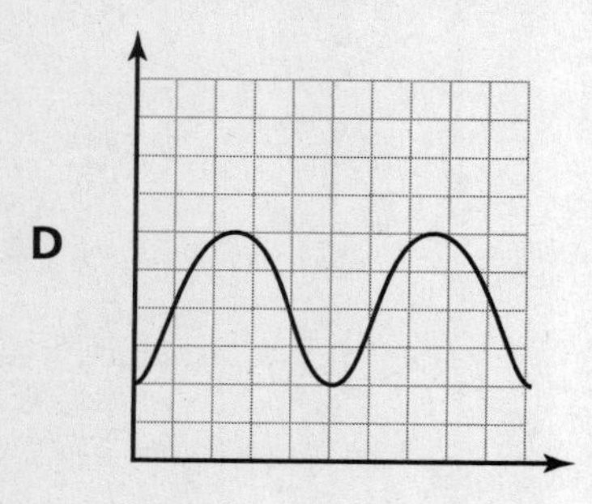

8 Factor the expression below.

$x^2 + 2x - 15$

A $(x - 3)(x - 5)$

B $(x - 3)(x + 5)$

C $(x + 3)(x - 5)$

D $(x - 15)(x + 1)$

9 In the diagram below, lines *j*, *k*, and *l* intersect.

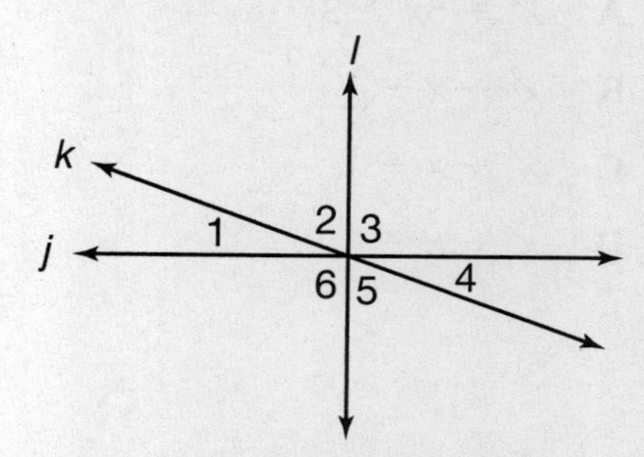

Which pair of angles have the same measure?

A $\angle 1$ and $\angle 2$

B $\angle 1$ and $\angle 3$

C $\angle 2$ and $\angle 5$

D $\angle 2$ and $\angle 6$

Go On

10 What is the slope of the line graphed below?

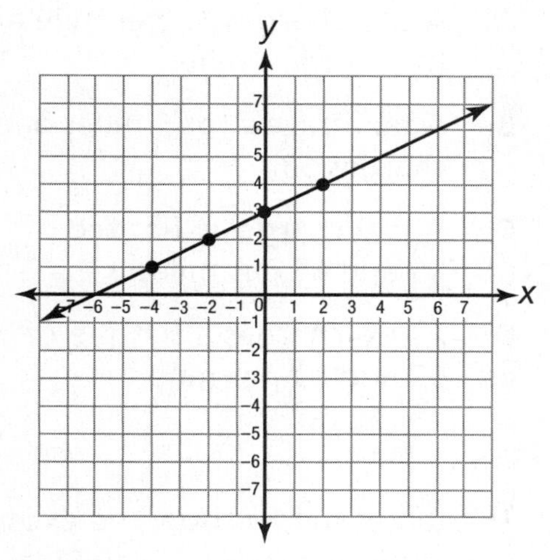

A -2

B $-\frac{1}{2}$

C $\frac{1}{2}$

D 2

11 Which of the following is a quadratic equation?

A $x^3 - 27 = 0$

B $x^2 - 27 = 0$

C $x - 27 = 0$

D $\frac{1}{x} - 27 = 0$

PRACTICE TEST 1

12  Use your ruler to help you solve this problem.

Last weekend, Lucy drove from Albany to Plattsburgh. Her route is shown on the map below.

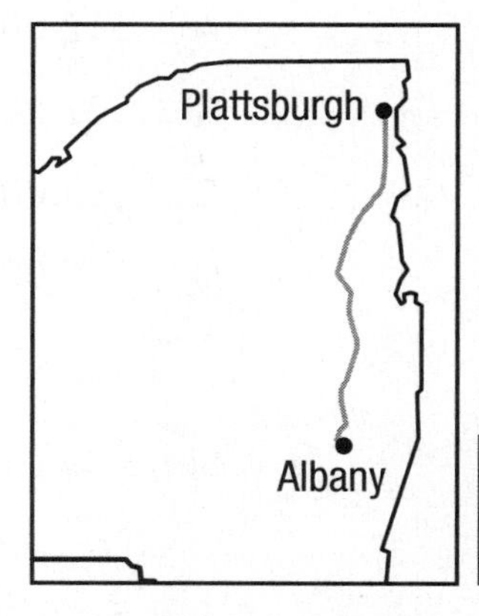

SCALE
$\frac{1}{2}$ inch = 60 miles

According to the map, what is the **approximate** distance Lucy drove?

A 100 miles

B 150 miles

C 175 miles

D 190 miles

Go On

13 An angle has a measure of 47°. What is the measure of the complement of this angle?

A 33°

B 43°

C 133°

D 143°

14 What is $4 \times 2^5 - 4^3$ in standard form?

A 28

B 64

C 192

D 32,704

15 Beth is planning to buy a printer with a price tag of $199. The sales tax is 8.375%. Which is the **best** estimate of the amount of tax Beth will pay?

A $8.00

B $10.00

C $15.00

D $20.00

16 Which equation represents the function with this table of values?

x	−2	−1	0	1	2
y	−3	−1	1	3	5

A $y = x - 1$

B $y = x + 1$

C $y = 2x - 1$

D $y = 2x + 1$

17 Which situation is **best** represented by the expression $5m + 3$?

A A taxi charges $5 per mile and travels for 3 miles.

B A taxi travels for 5 miles and charges $3.

C A taxi charges a flat fee of $3 and $5 per mile.

D A taxi charges a flat fee of $5 and $3 per mile.

18 The tables below show the exchange rate during Lester's visit to Scotland.

British Pound	U.S. Dollar
£1.00	$1.80

U.S. Dollar	British Pound
$1.00	£0.56

Lester paid 15 pounds for a theater ticket. How much was the ticket worth in U.S. dollars?

A $12.00

B $27.00

C $33.00

D $36.00

19 Simplify the expression below.

$$(3n^2 + 5n - 4) + (4n^2 + 1)$$

A $3n^2 + 9n - 3$

B $7n^2 + 5n - 3$

C $7n^2 + 5n - 4$

D $12n^2 + 5n - 5$

Go On

20 Simplify the expression below.

$$\frac{12a^5 - 8a^3}{4a^2}$$

A $3a^2 - 2a$

B $3a^3 - 2$

C $3a^3 - 2a$

D $3a^7 - 2a^5$

21 Factor the expression below using the greatest common factor (GCF).

$$15x^5 + 9x^3 + 12x$$

A $3x(5x^4 + 3x^2 + 4)$

B $3x(5x^5 + 3x^3 + 4x)$

C $3x(15x^5 + 9x^3 + 4)$

D $3x(5x^4 + 6x^2 + 4x)$

22 Which equation matches this graph?

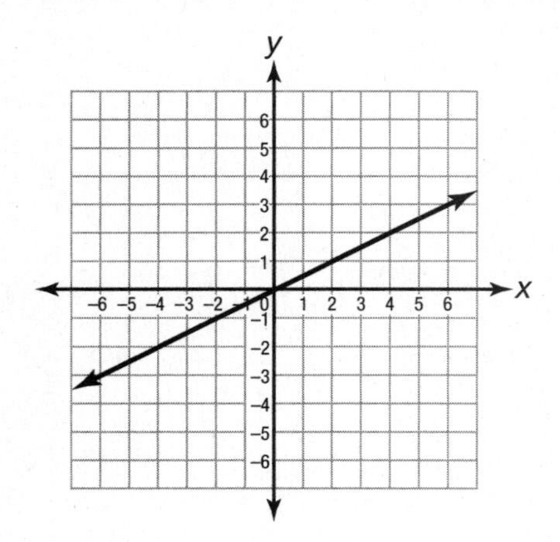

A $y = 2x$

B $y = x + 2$

C $y = \frac{1}{2}x$

D $y = x$

23 In the diagram below, lines l and m are parallel.

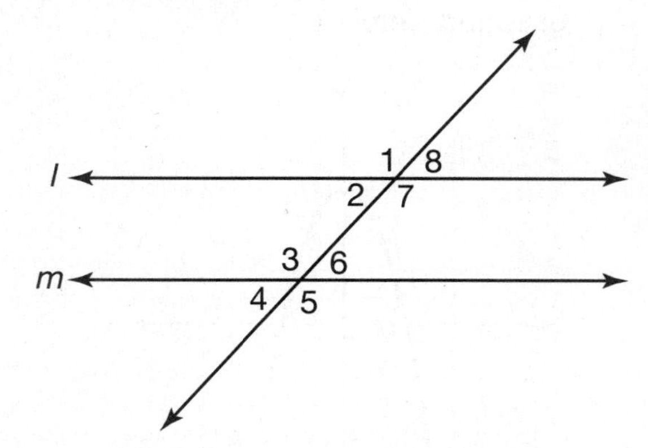

Which angle is supplementary to $\angle 6$?

A $\angle 2$

B $\angle 4$

C $\angle 7$

D $\angle 8$

24 A thermometer shows that the temperature is 5°C. What is the temperature in degrees Fahrenheit?

A −15°F

B 5°F

C 41°F

D 77°F

25 A line has a y-intercept of −8 and a slope of 4. What is the equation of the line in slope-intercept form?

A. $y = -8x + 4$

B. $y = 8x - 4$

C. $y = 4x + 8$

D. $y = 4x - 8$

Go On

26 Which of the following **best** describes the equation graphed below?

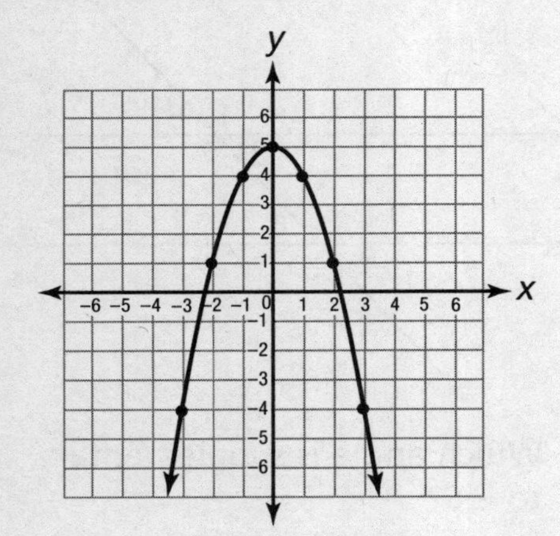

A nonlinear because the graph represents a straight line

B linear because the graph represents a curve

C nonlinear because the graph represents a curve

D linear because the graph represents a straight line

27 These are the prices of 4 brands of olive oil sold at a market.

Brand A: 12 ounces for $3.60

Brand B: 16 ounces for $5.40

Brand C: 10 ounces for $3.50

Brand D: 8 ounces for $3.20

Which brand is the **best** buy?

A Brand A

B Brand B

C Brand C

D Brand D

STOP

Session 2

28 In the diagram below, lines j and k are parallel.

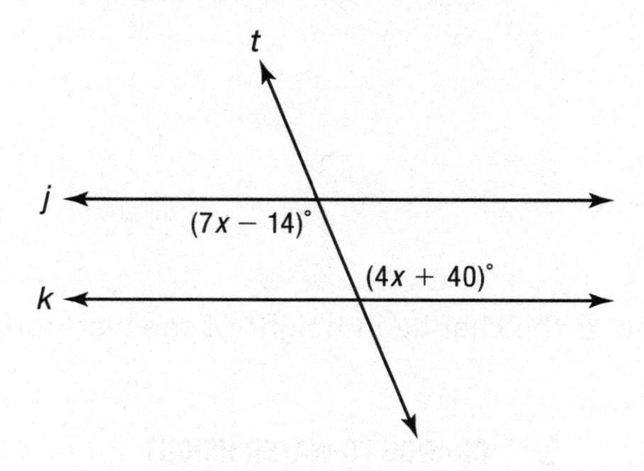

Part A

Solve for x.

Show your work.

Answer $x =$ ____________

Part B

What is the measure, in degrees, of the angle represented by $(7x - 14)$?

Answer ____________ degrees

Go On

29 Malcolm buys a denim jacket that is on sale for 15% off. The regular price is marked as $49.00. What is the sale price of the jacket?

Show your work.

Answer $____________

30 The graph below shows the change in height of the water in a swimming pool that is being drained.

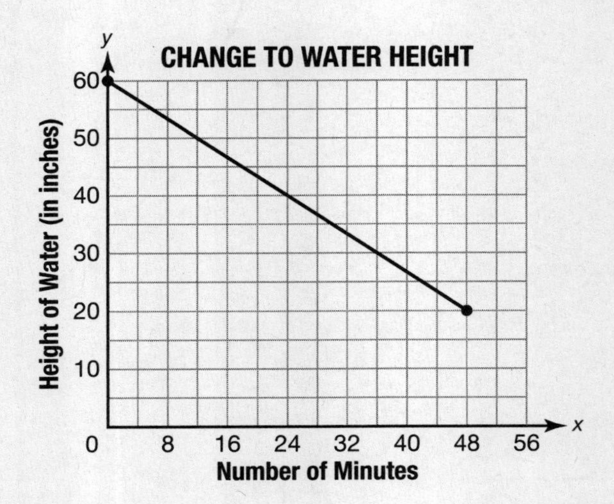

Use the information in the graph to determine the total minutes it takes for the height of the water to reach 0 inches.

Answer ____________ minutes

On the lines below, explain how you determined your answer.

____

____

____

____

____

Go On

31 Consider this inequality.

$2x + 5 \geq -1$

Part A

Solve the inequality.

Show your work.

Answer ______________

Part B

Graph your solution on the number line below.

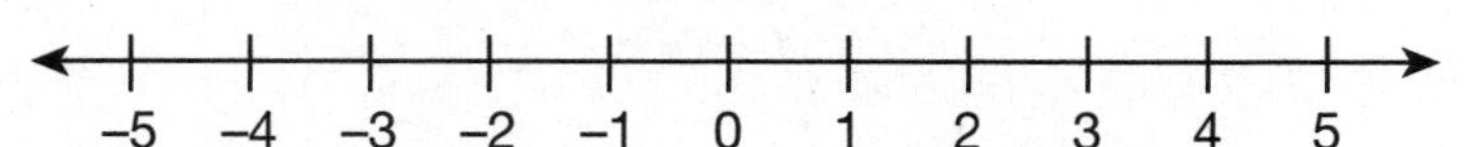

PRACTICE TEST 1

Go On

32 Gary multiplied the monomials $9a^2b^5$ and $6ab^3$ as shown below.

$(9a^2b^5)(6ab^3) = 54a^2b^8$

Is Gary's answer correct? On the lines below, explain how you determined your answer.

____

____

____

____

____

33 In the diagram below, lines *d* and *e* are parallel. The measure of $\angle 4$ is 74°.

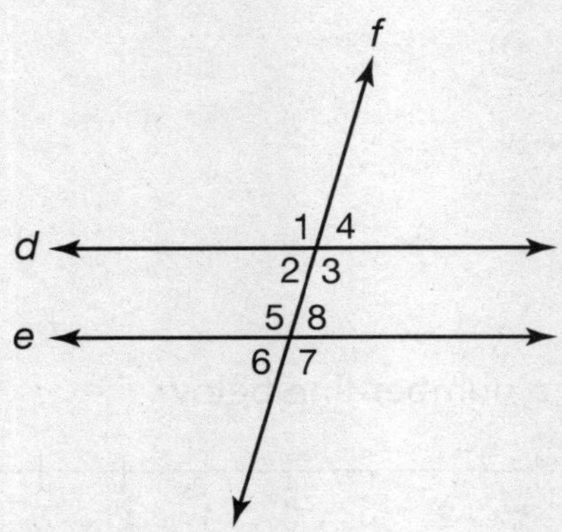

What is the measure of $\angle 5$?

Answer ____________________

STOP

Session 3

34 Graph this equation on the coordinate grid below.

$y = -\frac{2}{3}x - 2$

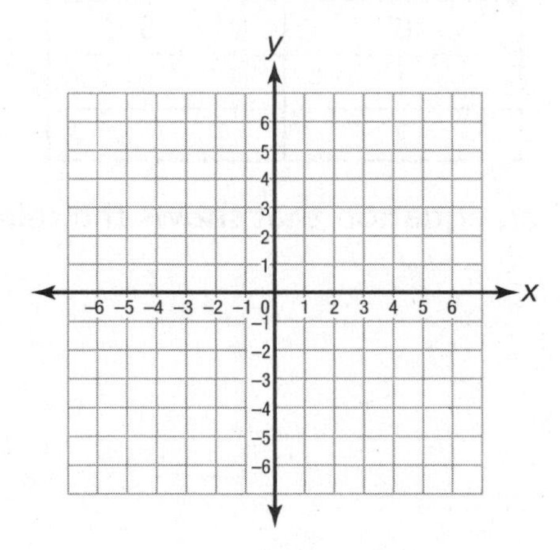

35 In the diagram below, lines *a* and *b* are parallel, and line *t* is a transversal.

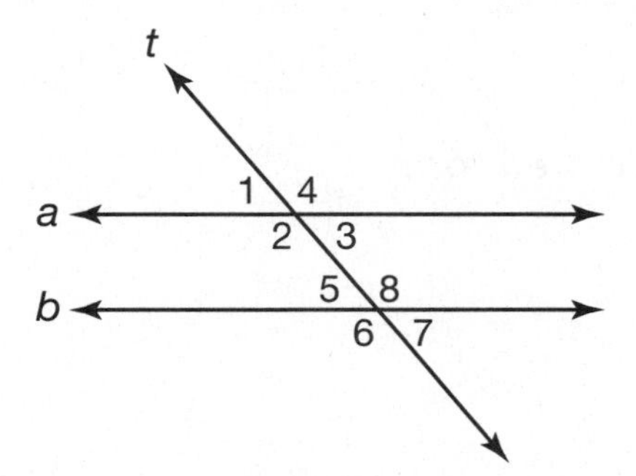

Name two angles in the diagram that are congruent to ∠5.

Answer ∠ ______________ and ∠ ______________

On the lines below, explain how you determined your answer.

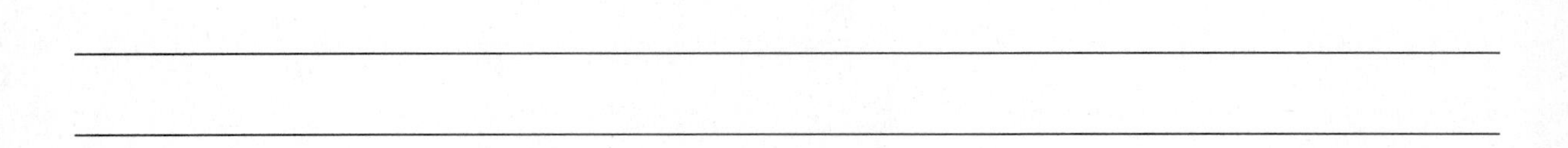

Go On

PRACTICE TEST 1

36 Complete the table below with the missing values for y.

x	y
-2	3
-1	4
0	5
1	
2	

On the line below, write an equation that shows the relationship between x and y in the table.

Answer ____________________

37 Write this trinomial in factored form.

$x^2 - 9x + 14$

Show your work.

Answer ____________________

Go On

38 Solve this inequality.

$-4x - 2 + x \leq x + 6$

Show your work.

Answer ______________

39 In the diagram below, lines WX and YZ intersect at point O. ∠WOY measures 28°. What is the measure of ∠YOX?

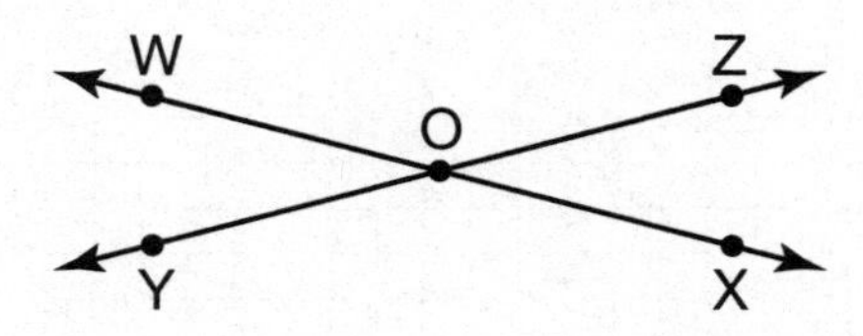

Show your work.

Answer ______________

Go On

40 The table below shows values for x and y when $y = 2x - 3$.

Part A

Complete the table by finding the value of y when $x = 2$.

x	−2	−1	0	1	2
y	−7	−5	−3	−1	

Part B

Plot the ordered pairs shown in the table on the coordinate plane below. Then draw a line connecting the points.

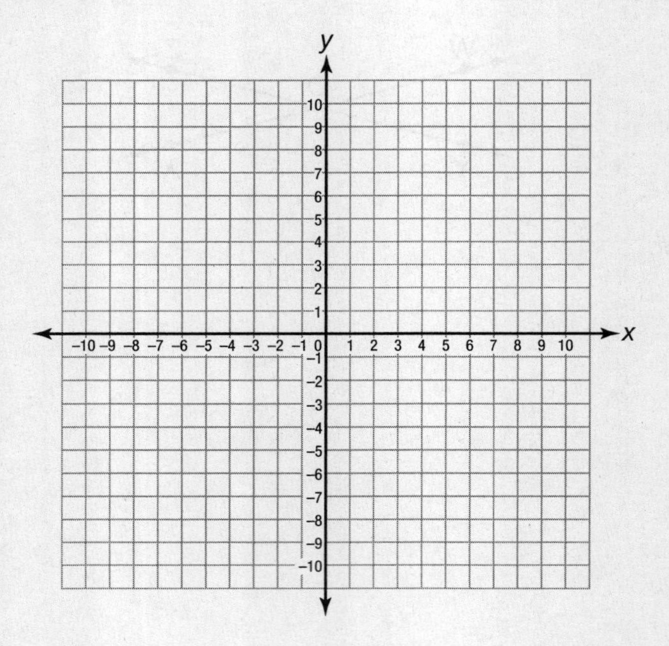

Part C

A point on the line has an x-coordinate of 3. What is the corresponding y-coordinate?

Answer ________________

Go On

41 Triangle RST and triangle R′S′T′ are plotted on the coordinate grid below.

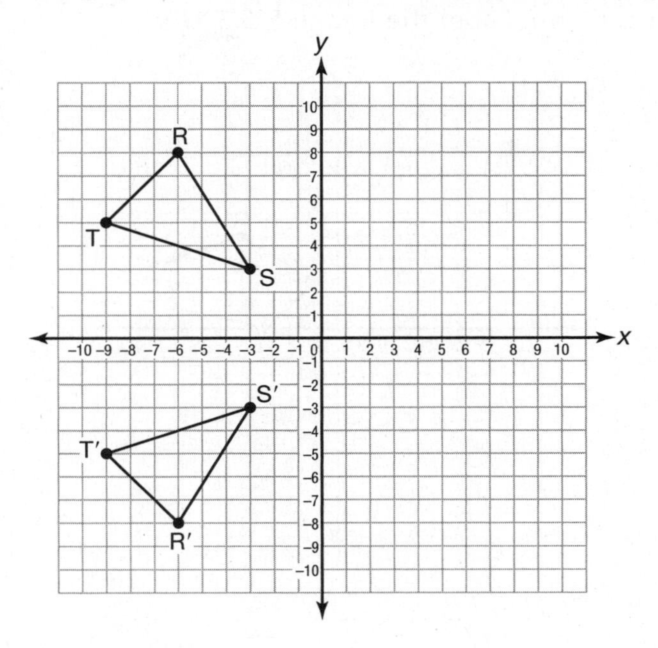

Part A

What is the name of the transformation applied to triangle RST that resulted in triangle R′S′T′?

Answer ____________________

Part B

On the lines below, describe how the coordinates of point R changed to the coordinates of point R′.

____

Go On

42 On the coordinate grid below, draw the image of polygon RSTUV translated 6 units to the left and 4 units down. Label the image R′S′T′U′V′.

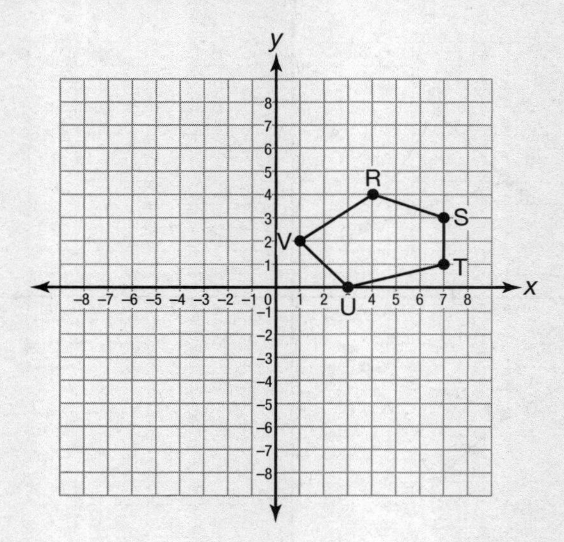

Go On

43 This is a system of linear equations.

$y = 2x + 6$

$y = -x + 3$

Part A

Graph this system of linear equations on the coordinate grid below.

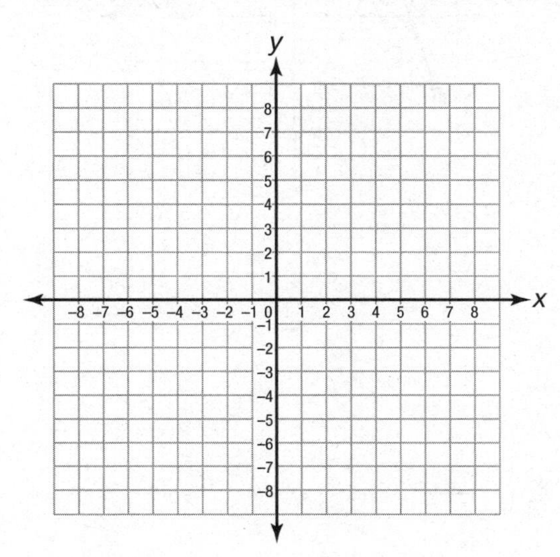

Part B

What is the solution of the system of linear equations?

Answer (____________, ____________)

Explain your solution on the lines below.

____

____

____

Go On

44 What are the coordinates of the vertices of triangle PQR after a reflection over the y-axis?

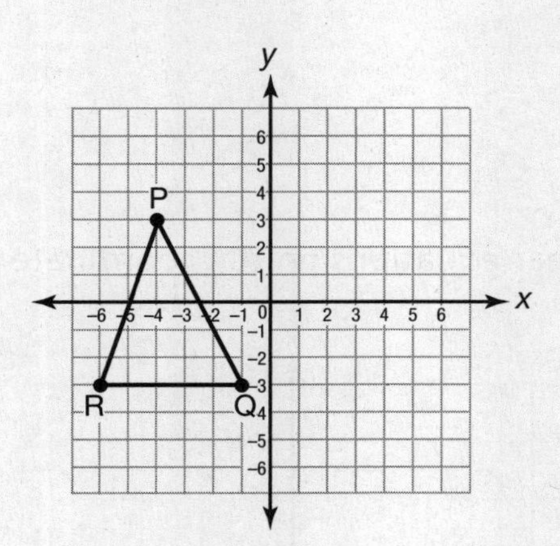

Show your work.

Answer

P′ __________ Q′ ____________ R′ ____________

45 What is the y-intercept of the line graphed below?

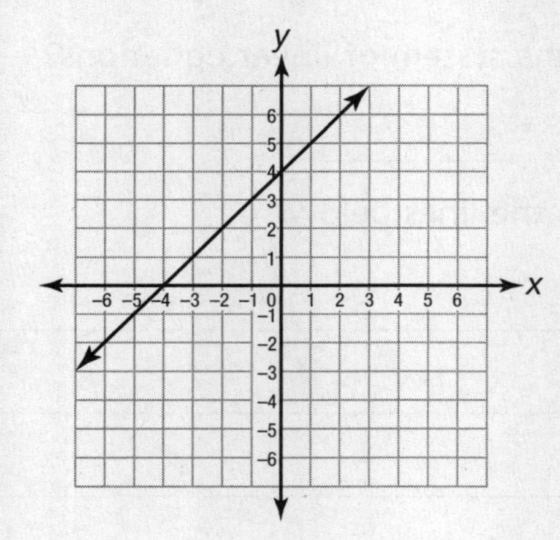

Answer _________________

STOP

New York State Coach, Empire Edition, Mathematics, Grade 8

PRACTICE TEST 2

Name: ______________________________

TIPS FOR TAKING THE TEST

Here are some suggestions to help you do your best:

- Be sure to read carefully all the directions in the test book.
- Read each question carefully and think about the answer before writing your response.
- Be sure to show your work when asked. You may receive partial credit if you have shown your work.
- Use your calculator to help you solve the problems in Sessions 2 and 3.

This picture means that you will use your ruler.

PRACTICE TEST 2

Session 1

1 What is $2^3 \times 2^4$ in exponential form?

A 2^{12}

B 2^7

C 2^1

D 2^{-1}

2 A company reported that its profits this year are 125% of last year's profits. What is 125% expressed as a decimal?

A 0.00125

B 0.125

C 1.25

D 12.5

3 There are 1,920 students in Lynn's high school. Twenty-nine percent of the students are freshmen. Lynn determined that there are about 557 freshmen. Which statement best explains whether or not Lynn's number is reasonable, and why?

A It is reasonable because $2{,}000 \times 0.3 = 600$.

B It is not reasonable because 20% of 1,900 is 380.

C It is not reasonable because $1{,}920 \div 30\%$ is 6,400.

D It is not reasonable because $1{,}920 \div 557$ is about 3.4.

4 Which equation represents the function with this table of values?

x	−2	−1	0	1	2
y	2	3	4	5	6

A $y = -x$

B $y = x + 4$

C $y = x - 4$

D $y = 2x + 4$

5 What verbal expression is the same as the algebraic expression below?

$9 - 5x$

A five times a number minus nine

B five minus nine times a number

C nine times a number minus five

D nine minus five times a number

6 What is the difference of the polynomials modeled below?

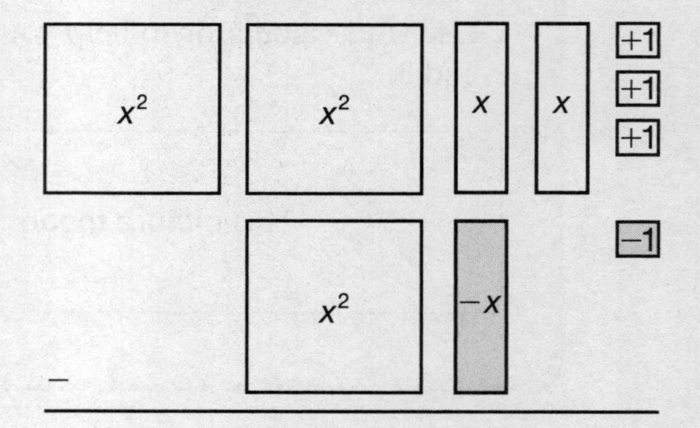

A $x^2 + 3x + 2$

B $x^2 + x + 2$

C $x^2 - 3x + 2$

D $3x^2 + 3x + 4$

Go On

7 Multiply.

$(n + 3)(4n - 1)$

A $4n^2 - 3$

B $4n^2 + 6n$

C $4n^2 + 11n - 3$

D $4n^2 + 13n - 3$

8 Factor the expression below.

$x^2 - 9x + 18$

A $(x + 3)(x + 6)$

B $(x - 3)(x - 6)$

C $(x - 2)(x - 9)$

D $(x + 2)(x - 9)$

9 In the diagram below, lines r, s, and t intersect.

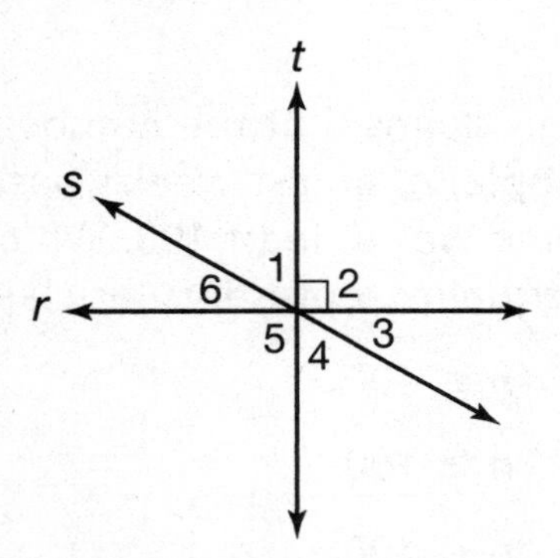

Which pair of angles have the same measure?

A $\angle 1$ and $\angle 2$

B $\angle 1$ and $\angle 4$

C $\angle 2$ and $\angle 4$

D $\angle 2$ and $\angle 6$

10 In the diagram below, $\angle AED$ measures $112°$.

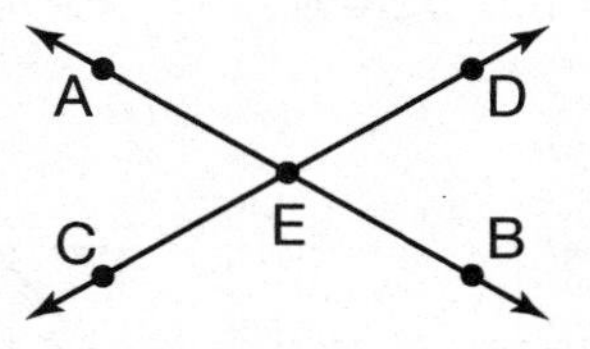

What is the measure of $\angle CEA$?

A $58°$ C $78°$

B $68°$ D $112°$

11 The area of rectangle PQRS is 72 square centimeters. Under which transformation could the area of the image, rectangle P′Q′R′S′, be less than 72 square centimeters?

A dilation C rotation

B reflection D translation

12 What is the solution for this inequality?

$3(x + 1) < 4x - 2$

A $x < 5$

B $x > 5$

C $x > 1$

D $x < -1$

PRACTICE TEST 2

Go On

13 Which of the following best describes the equation graphed below?

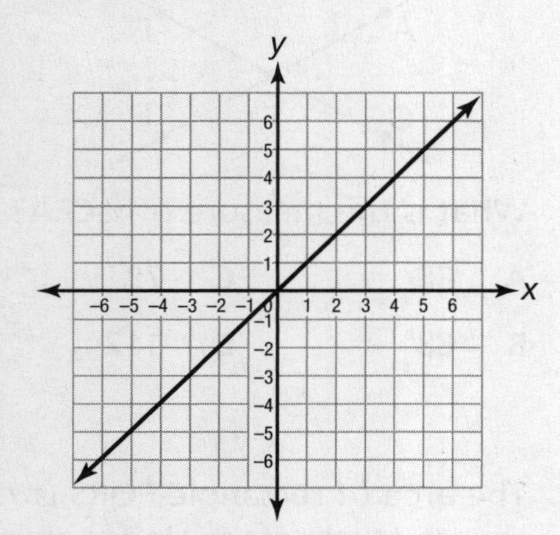

A nonlinear because the graph represents a straight line

B linear because the graph represents a curve

C nonlinear because the graph represents a curve

D linear because the graph represents a straight line

14 What is $4^3 \times 3^4$ in standard form?

A 144

B 1,728

C 5,184

D 823,543

15 Keisha took her family out to dinner. The restaurant check came to $97. She wants to leave a 20% tip. Which is the **best** estimate of the amount of the tip?

A $10.00

B $20.00

C $30.00

D $40.00

16 If a polygon has 5 sides, which expression can you use to find the sum of the measures of the interior angles of the polygon?

A $180(5 + 2)$

B $180(5 - 2)$

C $2(180 - 5)$

D $5(180 - 2)$

17 Juan claims that the number of people, *n*, at last week's baseball game was at least 100. Which inequality represents his claim?

A $n \leq 100$

B $n \geq 100$

C $n < 100$

D $n > 100$

Go On

18 Which graph best represents low-tide and high-tide water levels over time?

A
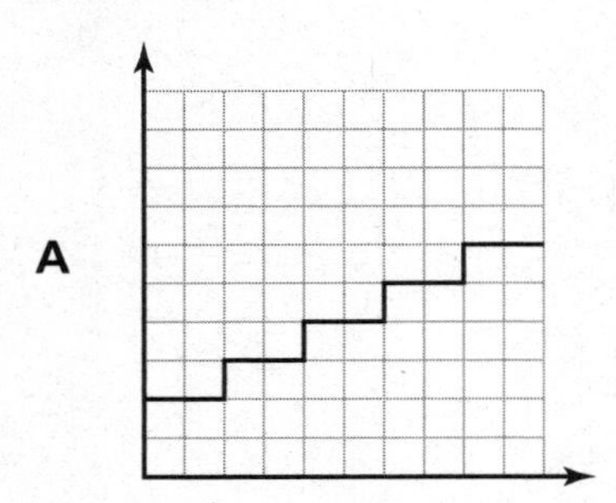

B
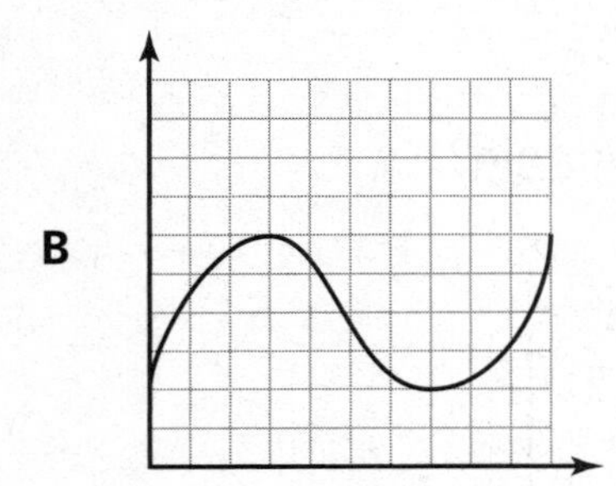

C
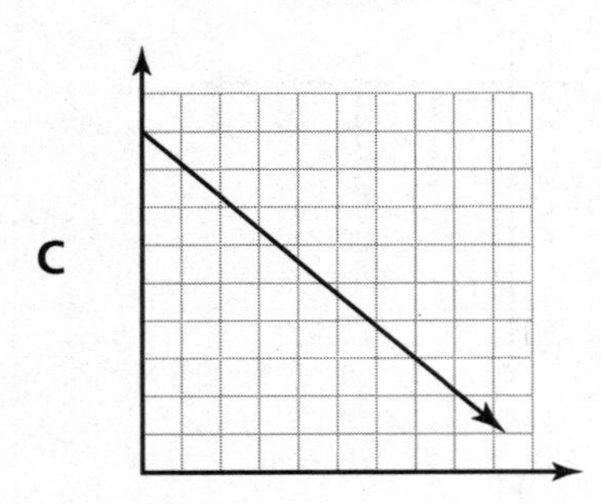

D
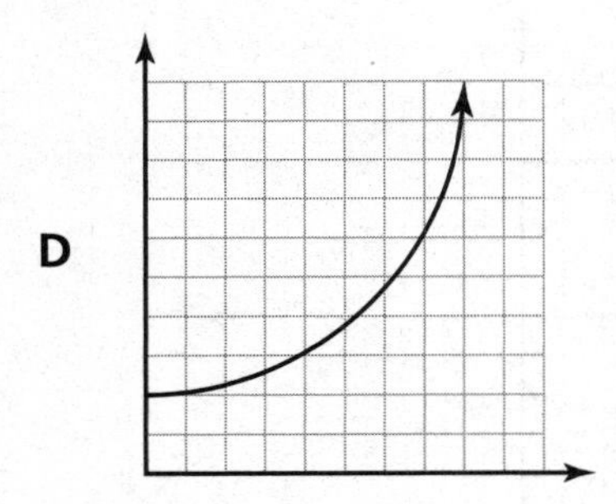

19 Simplify the expression below.

$$\frac{18x^3y^4}{3x^2y}$$

A $6xy^4$

B $6xy^3$

C $\frac{6y^3}{x}$

D $\frac{6}{xy^3}$

20 What is the simplified form of the expression below?

$$\frac{18n^6 - 12n^3}{3n^2}$$

A $6n^3 - 3$

B $6n^4 - 4$

C $6n^3 - 4n$

D $6n^4 - 4n$

21 Factor the expression below using the greatest common factor (GCF).

$$24n^6 + 16n^4 + 12n^2$$

A $4n^2(24n^6 + 16n^4 + 12n^2)$

B $4n^2(6n^6 + 4n^4 + 3n^2)$

C $4n^2(6n^4 + 4n^2 + 3)$

D $4n^2(6n^2 + 4n + 3)$

Go On

22 A rectangle is plotted on the coordinate plane below.

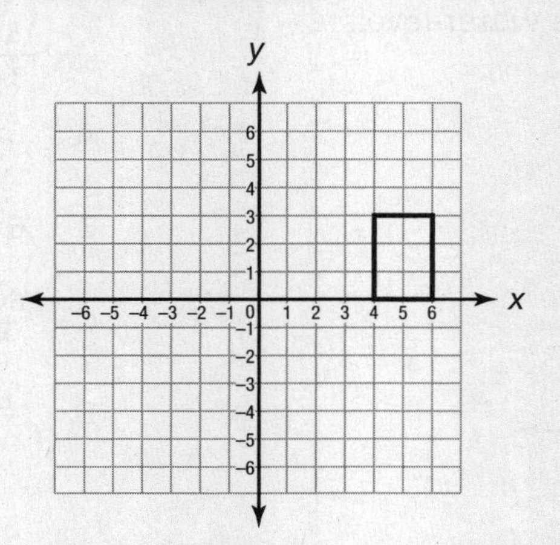

Which image shows a 90° clockwise rotation about the origin?

A

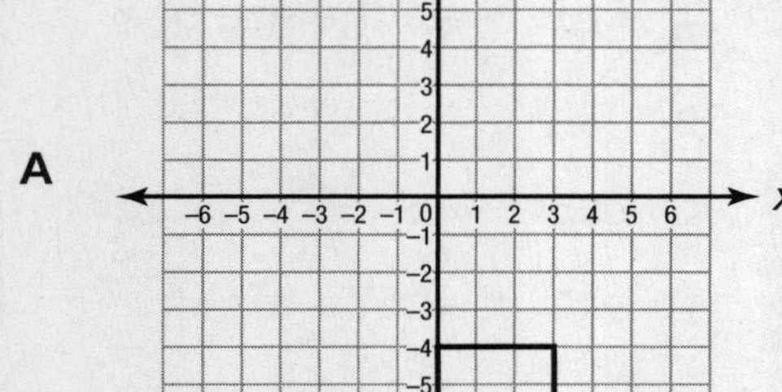

C

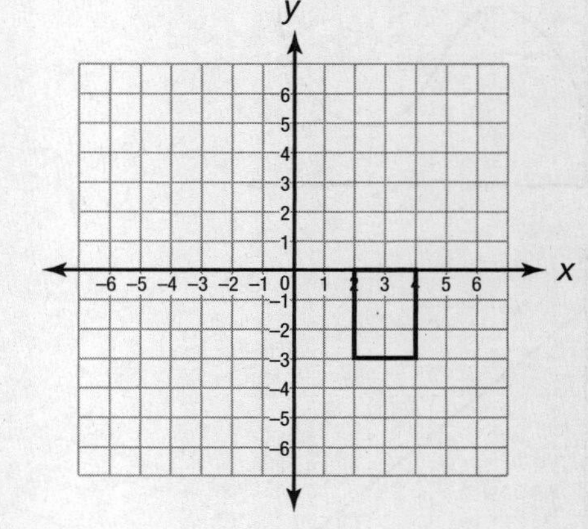

B

D

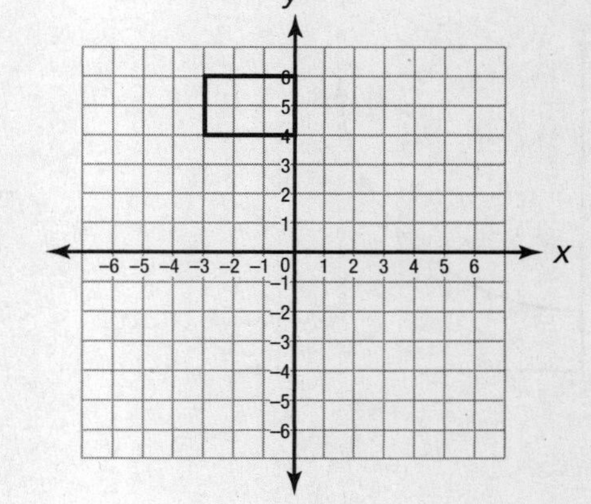

Go On

23 What is the *y*-intercept of this line?

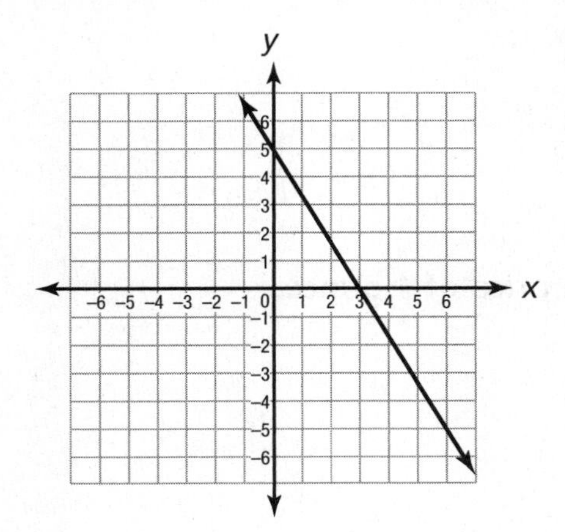

A $\frac{3}{5}$ C 3

B $\frac{5}{3}$ D 5

24 A thermometer shows that the temperature is 59°F. What is the temperature in degrees Celsius?

A 15°C C 48°C

B 18°C D 59°C

25 In the diagram below, lines *r* and *s* are parallel.

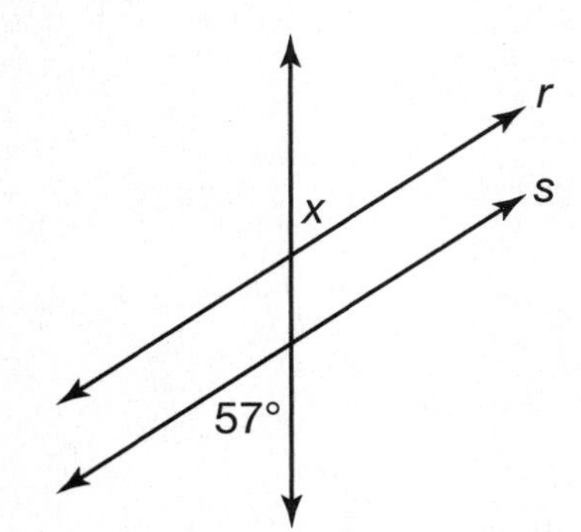

What is the measure of angle *x*?

A 33° C 57°

B 43° D 123°

26 Which statement matches this graph?

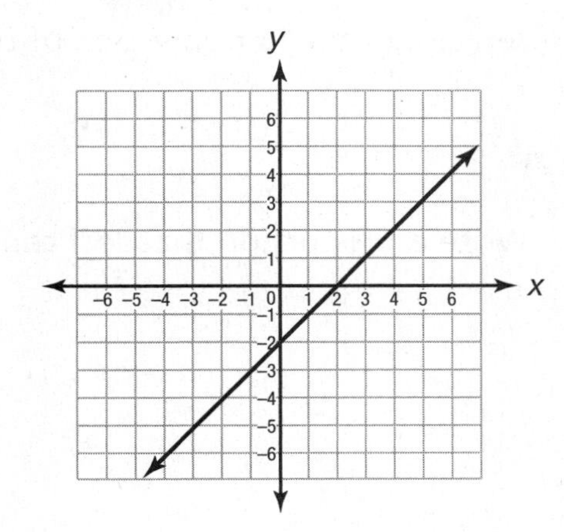

A *y* is 2 less than *x*.

B *y* is 2 more than *x*.

C *y* is twice the size of *x*.

D *y* is half the size of *x*.

27 What is the slope of the line graphed below?

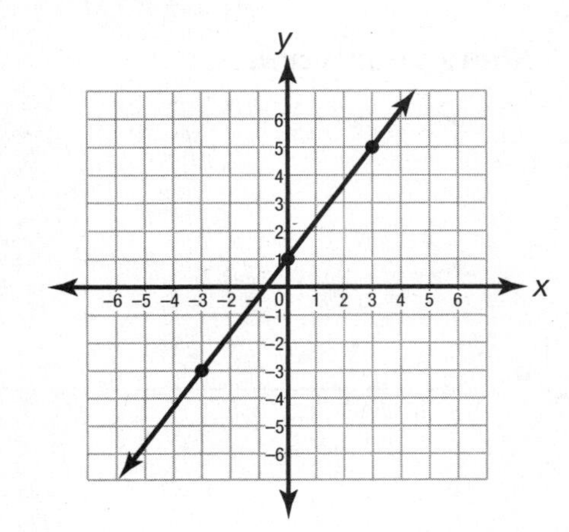

A $\frac{3}{4}$

B 1

C $\frac{4}{3}$

D 4

STOP

Session 2

28 Jeff bought a 16-ounce box of cereal for $4.80.

Part A

Write a proportion that Jeff can use to find the price of 1 ounce.

Answer ______________

Part B

Use your proportion to find the price of 1 ounce.

Show your work.

Answer $ ______________

Go On

29 The graph below shows how the temperature rose on a summer day.

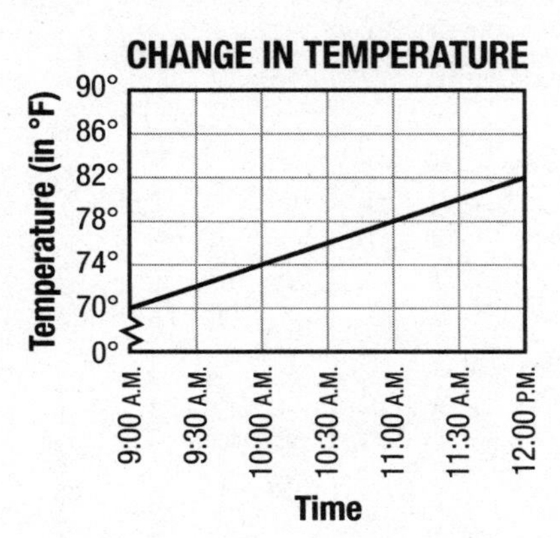

Use the information in the graph to determine the time at which the temperature reached 88°F.

Answer ________________ P.M.

On the lines below, explain how you determined your answer.

____

____

____

____

30 On a recent trip to Spain, Jodi used the exchange rates in the tables shown below.

U.S. Dollar	Euro
$1	€0.702247

Euro	U.S. Dollar
€1	1.424

What is the value of 75 U.S. dollars in euros? Round your answer to the nearest euro.

Show your work.

Answer € ______________

Go On

31 Consider this inequality.

$3x - 8 < 4$

Part A

Solve the inequality.

Show your work.

Answer ______________

Part B

Graph your solution on the number line below.

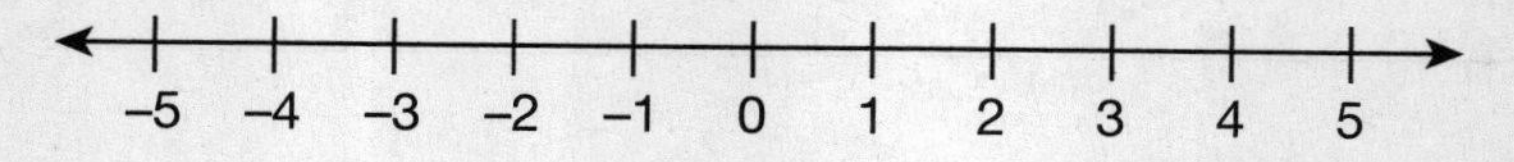

32 Simplify the expression below.

$(5x^2 + 3x - 8) - (x^2 - x + 5)$

Show your work.

Answer ______________

Go On

33 In the diagram below, lines *l* and *m* are parallel. Line *t* is a transversal.

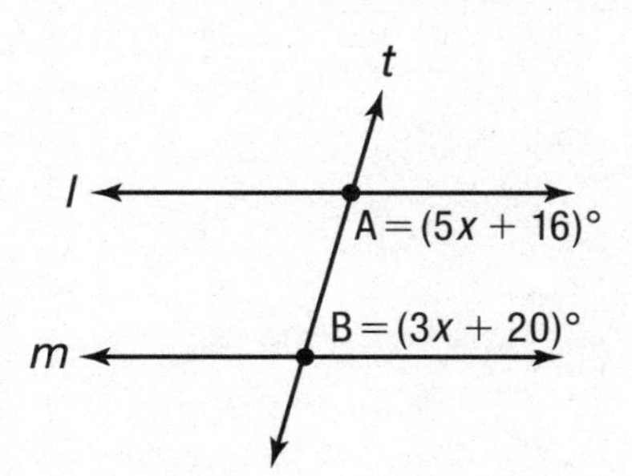

What is the measure, in degrees, of $\angle A$?

Show your work.

Answer ____________ degrees

STOP

Session 3

PRACTICE TEST 2

34 Solve this inequality.

$2(x - 3) > 3x - 4$

Show your work.

Answer ______________

35 In the diagram, what is the measure of angle x?

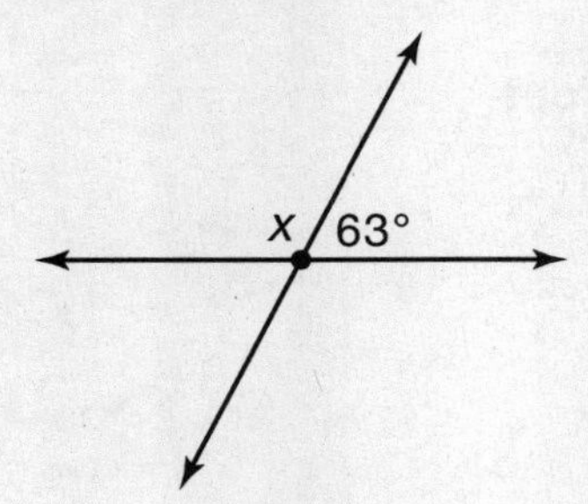

Show your work.

Answer ______________°

36 Write this trinomial in factored form.

$x^2 + x - 30$

Answer ______________

Go On

37 The Miller family bought 750 yards of rope for their new boat.

How many feet of rope did they buy?

Answer _______________ feet

38 Write the equation of the line graphed below.

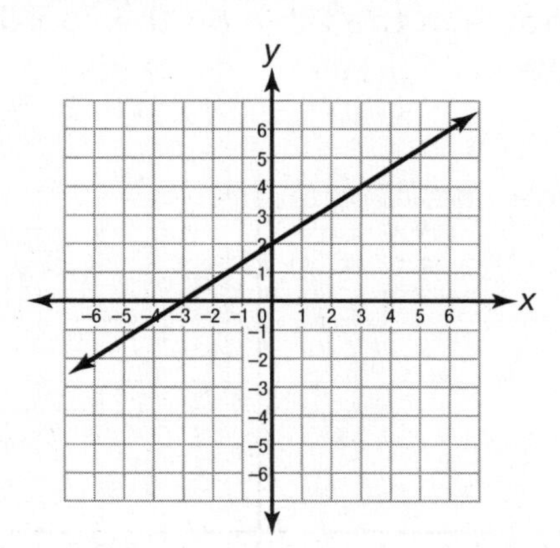

Answer y = _____________

39 On the coordinate plane, draw the image of triangle ABC after a reflection over the *y*-axis. Label the image A′B′C′.

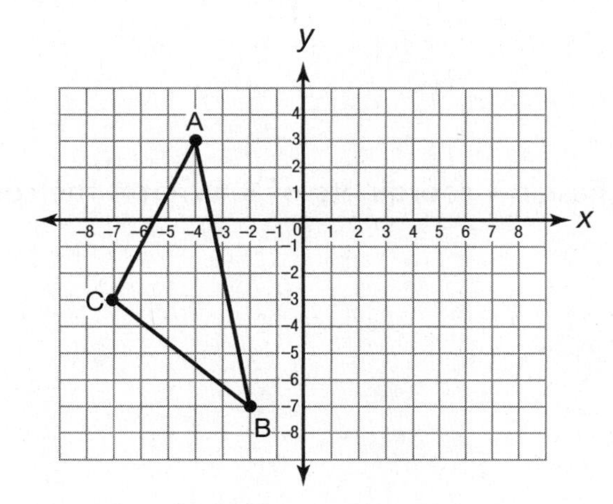

Go On

40 The table below shows values for x and y when $y = 3x - 3$.

Part A

Complete the table by finding the value of y when $x = 2$.

x	−2	−1	0	1	2
y	−9	−6	−3	0	

Part B

Plot the ordered pairs shown in the table on the coordinate plane below. Then draw a line connecting the points.

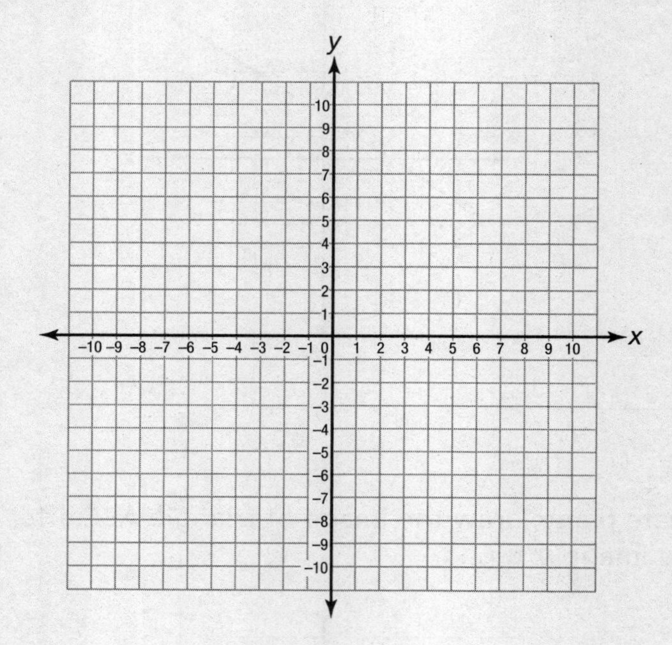

Part C

A point on the line has an x-coordinate of 3. What is the corresponding y-coordinate?

Answer ______________

Go On

41 The table below shows the coordinates of triangle DEF and the coordinates of D′ in triangle D′E′F′. Triangle D′E′F′ is a dilation of triangle DEF.

Triangle DEF		Triangle D′E′F′	
D	(−1, 2)	D′	(−2, 4)
E	(3, −3)	E′	
F	(−2, −2)	F′	

Part A

What are the coordinates of point E′ and point F′?

Answer E′ = (________ , ________)

F′ = (________ , ________)

Part B

On the grid below, draw triangle DEF and triangle D′E′F′.

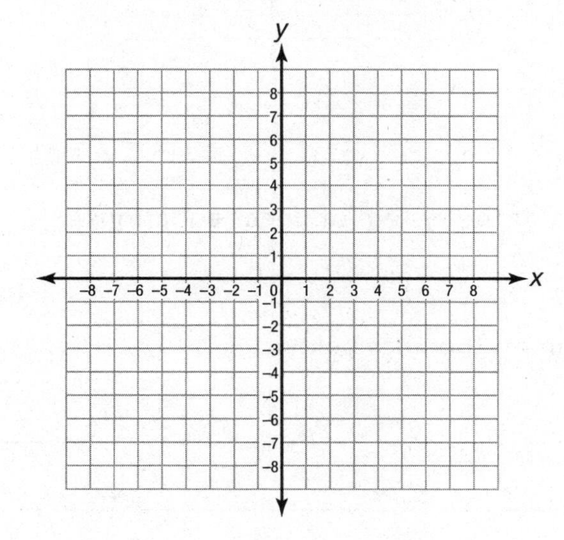

Go On

42 This is a system of linear equations.

$y = 4x + 1$

$y = -2x + 7$

Part A

Graph this system of linear equations on the coordinate grid below.

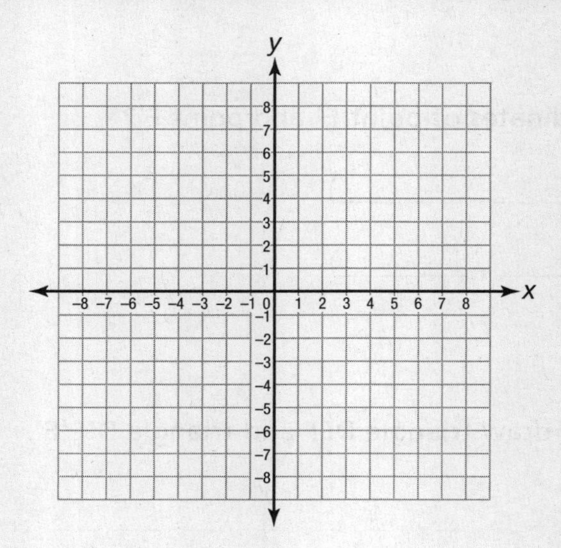

Part B

What is the solution of the system of linear equations?

Answer (________ , ________)

Explain your solution on the lines below.

____

____

____

Go On

43 On the coordinate plane below, draw the image of parallelogram JKLM translated 5 units to the right and 8 units down. Label the image J′K′L′M′.

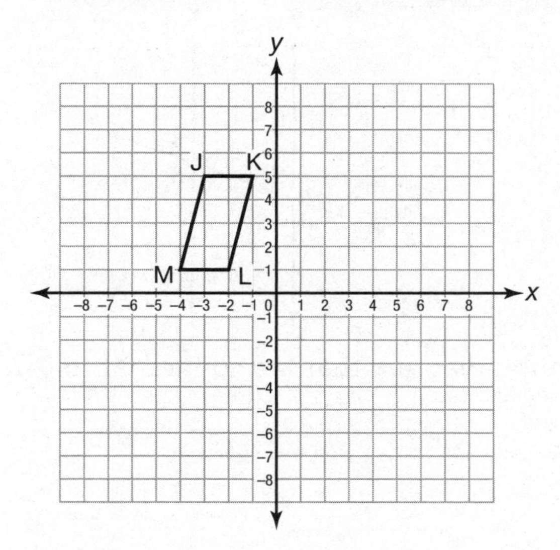

Go On

44 In the diagram below, lines *x* and *y* are parallel and line *z* is a transversal.

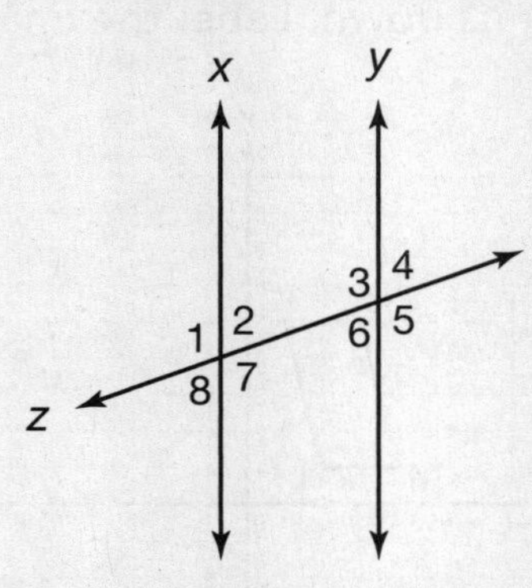

Name two angles in the diagram that are congruent to ∠7.

Answer ∠ ______________ ***and*** ∠ ______________

45 In the diagram below, lines *a*, *b*, and *c* intersect as shown. Lines *a* and *c* are perpendicular. ∠3 is complementary to ∠2.

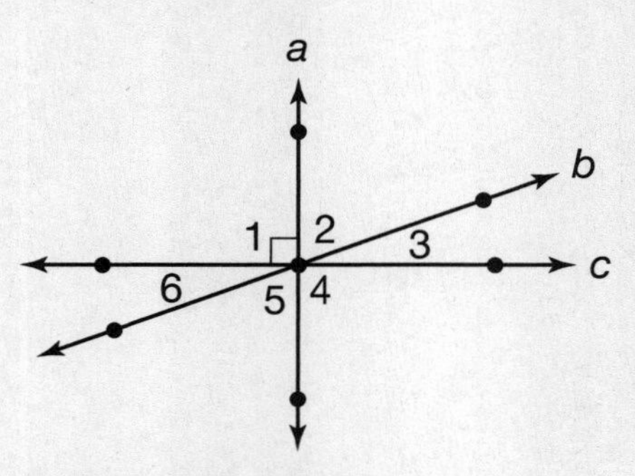

What other angle is complementary to ∠2?

Answer ∠ ________________

STOP

Mathematics Reference Sheet

FORMULAS

Pythagorean Theorem

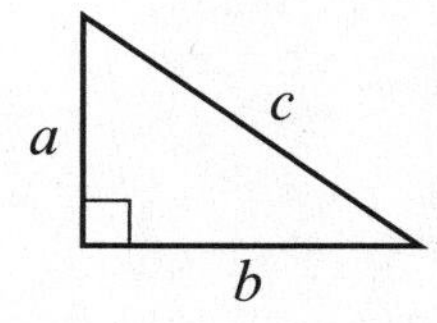

$c^2 = a^2 + b^2$

Simple Interest $I = prt$

Distance Formula $d = rt$

CONVERSIONS

Temperature Conversions

$F = \frac{9}{5}C + 32$

$C = \frac{5}{9}(F - 32)$

Measurement Conversions

1 mile = 5,280 feet
1 yard = 3 feet

Punch-Out Tools

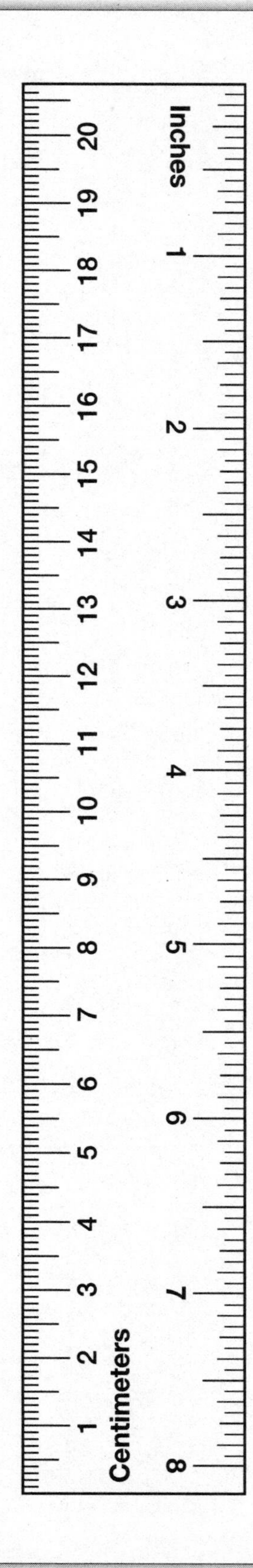

Notes